Afghanistan–Pakistan Shared Waters

Afghanistan–Pakistan Shared Waters

State of the Basins

Edited by

Muhammad Azeem Ali Shah

International Water Management Institute, Pakistan

Jonathan Lautze

International Water Management Institute, South Africa

Asadullah Meelad

Freelance Consultant

CABI

USAID
FROM THE AMERICAN PEOPLE

IWMI
International Water
Management Institute

CABI is a trading name of CAB International

CABI
Nosworthy Way
Wallingford
Oxfordshire OX10 8DE
UK

Tel: +44 (0)1491 832111
E-mail: info@cabi.org
Website: www.cabi.org

CABI
200 Portland Street
Boston
MA 02114
USA

T: +1 (617)682-9015
E-mail: cabi-nao@cabi.org

A catalogue record for this book is available from the British Library, London, UK.

A catalogue record for this book is available from the British Library, London, UK.

Library of Congress Control Number: 2023933433

ISBN-13: 9781800622357 (hardback)
9781800622364 (OA ePDF)
9781800622371 (OA ePub)

DOI: 10.1079/9781800622371.0000

Commissioning Editor: David Hemming
Editorial Assistant: Lauren Davies
Production Editor: Marta Patiño

Typeset by Exeter Premedia Services Pvt Ltd, Chennai, India
Printed and bound in the USA by Integrated Books International, Dulles, Virginia

This book was made possible through support provided by the US Agency for International Development's Mission to Pakistan (USAID/Pakistan) under the terms of Award No. 72039118IO00003. The opinions expressed in this book are those of the authors and do not necessarily reflect the views of the US Government or US Agency for International Development.

Contents

Contributors

Taimoor Akhtar, University of Guelph, Canada; email: taimoor.akhtar@gmail.com

Arif Anwar, University of Birmingham, Pakistan; email: a.a.anwar@bham.ac.uk

Mujib Ahmad Azizi, Afghanistan Research and Evaluation Unit, Afghanistan; mujeebahmad.azizi@gmail.com

Muhammad Tousif Bhatti, International Water Management Institute, Pakistan; email: tousif.bhatti@cgiar.org

Marie-Charlotte Buisson, International Water Management Institute, Sri Lanka; email: m.buisson@cgiar.org

Muhammad Zia ur Rahman Hashmi, Global Change Impact Studies Centre, Pakistan; email: ziahashmi77@gmail.com

Shakeel Hayat, Institute of Management Sciences Peshawar, Pakistan; email: shakeel.hayat@imsciences.edu.pk

Muhammad Shahid Iqbal, Institute of Space Technology, Pakistan; email: shahid.iqbal@cgiar.org

Inayatullah Jan, University of Agriculture Peshawar, Pakistan; email: jaan.inayat@gmail.com

Hassaan Furqan Khan, Habib University, Pakistan; email: Hassaan.khan@sse.habib.edu.pk

Jonathan Lautze, International Water Management Institute, South Africa; email: J.Lautze@cgiar.org

Karthikeyan Matheswaran, International Water Management Institute, Sri Lanka; email: k.matheswaran@cgiar.org

Asadullah Meelad, Freelance Consultant; email: asadmeelad@gmail.com

Fazilda Nabeel, University of Sussex, UK; email: F.Nabeel@sussex.ac.uk

Paul Pavelic, International Water Management Institute, Lao PDR; email: p.pavelic@cgiar.org

Muhammad Azeem Ali Shah, International Water Management Institute, Pakistan; email: a.shah@cgiar.org

Alias Wardak, Universität Siegen, Germany; email: wardak@architektur.uni-siegen.de

The Editors

Muhammad Azeem Ali Shah is an international researcher working on water governance institutions at the International Water Management Institute, Lahore, Pakistan. His research focuses on water governance, institutions and policy analysis, transboundary water management and solar irrigation systems.

Jonathan Lautze is Research Group Leader, Integrated Management of Basins and Aquifers, at the International Water Management Institute in South Africa. His areas of expertise include water security, transboundary waters, the water–energy–food nexus and water health.

Asadullah Meelad is a freelance consultant based in Kabul. His areas of research include transboundary water co-operation, international water law and water governance. He served as Senior Legal and Policy Advisor on water issues to the government of Afghanistan.

Foreword

Afghanistan and Pakistan share nine river basins within the broader Indus river system. The resources of these basins provide intrinsic value – and pose some threats – to over 43 million people, about one third of the population in Afghanistan and most of the population of Khyber Pakhtunkhwa Province in Pakistan. Intrinsic value is evidenced through the essential contribution of water to basic human consumption and religious needs, as well as the role it plays in livelihoods and food security. Threats are manifested as floods, which routinely cause economic losses and occasionally loss of lives, increasing competition for a limited water resource and potential for conflict as river flows become more variable.

Conditions in Afghan–Pakistani shared waters are dynamic rather than static. Basin realities are changing rapidly as the population grows, the political context evolves, and the effects of climate change become ever more evident. While responding to current and future conditions requires a cooperative basin approach, there is as yet no water agreement on any of the shared rivers. Both countries have followed a unilateral pathway in developing the water resources in their portion of these river basins. The absence of a basic level of co-ordination represents a lost opportunity to advance the sustainable development aims pursued by both countries and an equitable outcome.

While advancing co-operation on several natural resources is often a 'long haul' that requires efforts on multiple fronts, and a certain level of opportunism to identify elements of mutual interest, a starting point is to foster a common understanding of the shared resource. Stated simply, it is difficult to jointly manage a shared watercourse if there are divergent views on what is being managed. While there have been efforts in the past to carry out research on "water resources within individual countries," there has been little effort to compile, integrate and interpret data and information from both sides, and a comprehensive assessment of the state of these river basins is yet to be developed.

This book strives to fill this gap by providing a comprehensive baseline assessment of three major basins shared by Afghanistan and Pakistan: the Kabul, Kurram and Gomal. Toward this end, the authors have provided evidence-based knowledge and information on socio-economic development, climate change impacts, surface and groundwater conditions, land use and land cover changes, water governance and institutional landscape. While compiling data is useful, they also want to emphasize that data should be a means to something bigger: co-operation and joint development of the shared water resources between Afghanistan and Pakistan.

A critically important feature of this book is the co-development approach through which chapters have been drafted, which is jointly by authors from Afghanistan and Pakistan with support from international experts from the International Water Management Institute and funding support from

USAID. This collaborative approach helped ensure a factual and balanced view across the chapter topics with local insights provided through access to an important body of grey literature.

The findings in this book underscore the enormous untapped potential of the three river basins, which, if developed through cooperative mechanisms, could bring millions of people out of extreme poverty. A key to such development will be transboundary water co-operation between Afghanistan and Pakistan supported by re-engagement with the international community to create a context that attracts investment. The arrival of a new government in Afghanistan poses new challenges and delays related to international recognition and formulating new government structures. At the same time, unforeseen opportunities could emerge as the new government may foster better relations with Pakistan.

I welcome this first comprehensive and multidisciplinary book on the state of the Kabul, Kurram and Gomal river basins shared between Afghanistan and Pakistan. It provides a strong knowledge base for policymakers, practitioners, academics, students and readers who wish to acquaint themselves with the current state of affairs on the shared river basins between Afghanistan and Pakistan. Stakeholders in both Afghanistan and Pakistan, development partners and others are encouraged to consider its recommendations for strengthening our knowledge of these basins as a starting point for future collaboration.

Jeremy D. Bird
Former CEO Mekong River Commission Secretariat
Former DG, International Water Management Institute
Date: June 2023

Preface

In June 2022, Vice President Kamala Harris launched the *White House Action Plan on Global Water Security*, a landmark whole-of-government effort to achieve a water-secure world. The plan covers the full spectrum of global water issues and advances United States leadership on water security. The problem statement notes that water security is under increasing threat worldwide and that water insecurity is growing.

The plan outlines key links between water insecurity and shared watercourses. Specifically identified is: 'increasing risk that actors controlling the headwaters of key rivers or water infrastructure will monopolize these resources, either to supply domestic needs or to exert pressure on downstream countries'. Equally, the plan notes that water management can directly influence political outcomes and 'cooperation on transboundary water resources has been shown to reduce the risk of inter-state conflict'. Ultimately, the plan calls for a systems approach that transcends political boundaries to address the growing global water insecurity problem and ensure that water strengthens, rather than undermines, US national security.

In the context of water management, realizing the ambitious aim of a systems-based approach across political boundaries typically means conforming to the boundaries of river basins. As noted in Pillar 2 of the White House Action Plan (Promoting sustainable management and protection of water resources and associated ecosystems to support economic growth, build resilience, mitigate the risk of instability or conflict, and increase co-operation), collaborative monitoring and data collection and analysis will be the key to scoping the opportunities through which systems approaches can be pursued. Likewise, Pillar 2 outlines the need for the US government to support agreements among stakeholders sharing water resources to reduce conflict and contribute to broader water security aims.

This book supports the realization of Pillar 2 of the *White House Action Plan on Global Water Security*. The editorial and author teams have focused on various aspects of water management typically needed to pursue a systems approach in three critical river basins shared by Afghanistan and Pakistan (the Kabul, Kurram and Gomal). The book project adopted a 'co-production approach' which helped foster the convergence in thinking that forms the foundation for negotiated cooperative water agreements that reduce conflict and enhance water security. Above and beyond these contributions, we believe this to be one of the first efforts to advance knowledge on the three shared basins by integrating data from the two countries.

In conclusion, we hope this book reflects the beginning rather than a conclusion to the process. The history of Afghanistan and Pakistan may present one of the more challenging contexts in which to promote water cooperation and may discourage some from engaging here. I believe these greater challenges are precisely why we must engage. In the context of Afghanistan and Pakistan, cooperation on water between the two countries can provide direct benefits to water-dependent sectors that contribute to sustainable development. But cooperation on water can also provide a range of indirect benefits such as strengthening trust, fostering regional interaction that leads to benefits in sectors far removed from water, and reducing the risk of conflict.

Muhammad Azeem Ali Shah
Jonathan Lautze
Asadullah Meelad
Date: June 2023

1 Introduction

Muhammad Azeem Ali Shah, Jonathan Lautze and Asadullah Meelad

Chapter Overview

This chapter contextualizes the book's aim and scope within global experiences in transboundary water management as well as regional realities of human development. The chapter rationalizes the focus on three of the nine rivers shared by Afghanistan and Pakistan, i.e., the Kabul, Kurram and Gomal, and identifies the new contribution that this book provides. Importantly, this is believed to be the first book to focus on the entirety of these three shared rivers—compiling and synthesizing data and information from both countries to distil key lessons and messages. The chapter proceeds to provide a cursory overview of the three basins to lay foundations for the more thorough elucidations in the chapters that follow. The chapter concludes with reflections on the significance of this book and its role in ameliorating conditions in the three basins.

Fig. 1.1. Kabul river near Amangarh Nowshera, Khyber Pakhtunkhwa, Pakistan. (Used with permission). Photo credits: Amjad Jamal - IWMI.

DOI: 10.1079/9781800622371.0001

International River Basins

Freshwater resources remain a globally threatened renewable resource despite being fundamental to an extensive range of essential ecological and societal activities, from agriculture to hydropower generation, industrial expansion and waste disposal. Freshwater resources are bounded within river basins – physiographic units circumscribed by topographic divides that limit the areas of land drained by a main river, its affluents and sub-affluents (Arai *et al.*, 2012). River basins are often distributed across political, ethnic and cultural borders, which can give rise to intense hydro-political disputes.

Over 260 transboundary river basins cover 45% of the global land surface and account for 60% of the overall global river flow (Giordano *et al.*, 2014; Wolf *et al.*, 2003). The borders that give rise to international waters thus constitute a formidable factor that confounds optimal basin management and holds the potential to aggravate interstate tensions in response to concerns of perceived inequities in the allocation, use and management of water resources. To address this, three international directives offer guidance on co-operation in international watercourses. The 1992 United Nations Economic Commission for Europe (UNECE) Water Convention serves as a mechanism to provide the countries sharing transboundary watercourses with a way to develop bilateral or multilateral agreements to prevent, control and reduce transboundary water impacts (Article 9). It also stipulates (Articles 3 and 8) that neighbouring states 'may enter into, or consider harmonizing, existing watercourse agreements with the basic principles of the Convention and may consider the establishment or joint mechanisms or commissions' (United Nations, 2022). Similarly, the 1997 UN Convention on the Law of Non-Navigational Uses of International Watercourses calls for and guides countries in the application of equitable and sustainable use and 'no significant harm' (Votrin, 2005). Finally, the 2004 Berlin Rules build on existing conventions but aim to better integrate emerging principles of ecological integrity, sustainability and public participation, also incorporating the core fundamental international environmental and human rights (Salman, 2007; Cooley *et al.*, 2009).

There are numerous examples of successful co-operation across transboundary watercourses including the Rhine, Columbia, Danube, Senegal, Orange-Senqu, Niger and Parana rivers (Iqbal, 2020). Some of the prominent features of this co-operation include joint planning and decision-making, respect for international water laws, economic collaboration, a shared knowledge base and data exchange, involvement of key stakeholders who are not limited to government entities, and joint planning in the face of uncertainties (United Nations, 2022).

Success in the present may not necessarily withstand intensifying pressures, however. For instance, increased population, interlinked economic growth and, most important, the already changing climate may test the suitability of some international water agreements (De Stefano *et al.*, 2012). If we cannot address these pressures, we will clearly jeopardize the current co-operative interstate relations of transboundary basin states.

Afghanistan–Pakistan Shared Waters

Afghanistan and Pakistan share nine rivers with a combined population of more than 43 million (Fig. 1.2). These watersheds have annual flows of about 26 billion m^3 (bcm) per year (details in Chapter 3). A growing population is driving increased demand for resources on both sides of the border. Afghanistan occupies an unusual position in which 90% of its surface water resources are shared with downstream neighbouring countries, which include Pakistan, Iran, Tajikistan and Turkmenistan (Thomas *et al.*, 2016). Afghanistan has experienced instability and conflict, which has led to an impasse in hydrological infrastructure development. After the regime change in 2001, Afghanistan developed ambitious plans to build dams and canal infrastructure. This resulted in objections from Iran, Pakistan and Turkmenistan (Iqbal, 2020). Not surprisingly, the unilateral development of transboundary water resources in Afghanistan has attracted little support from multilateral institutions.

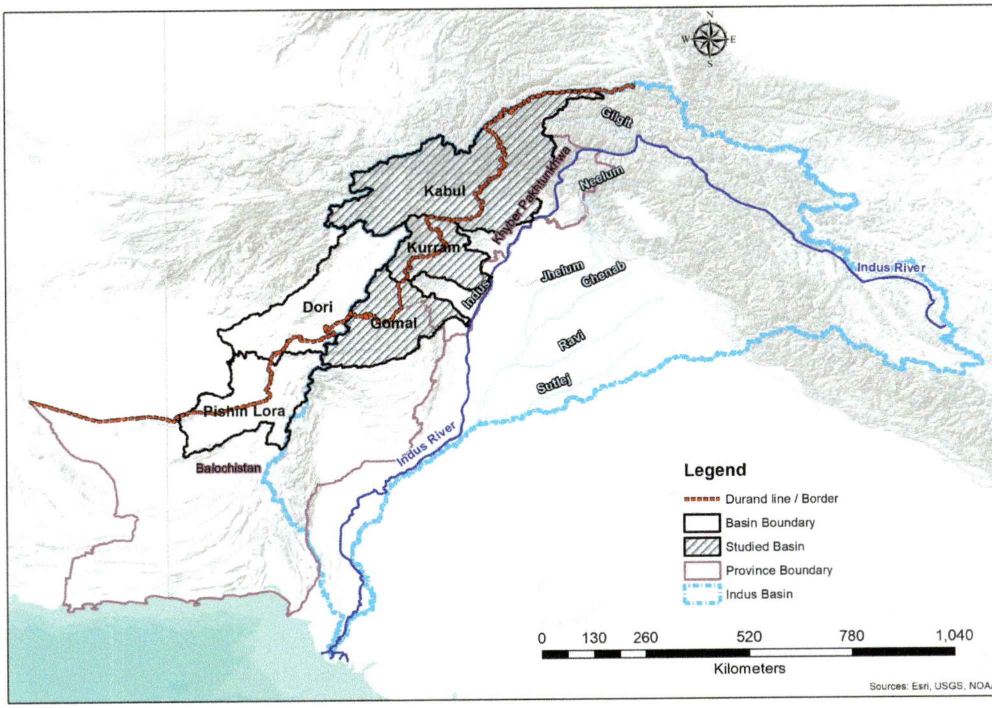

Fig. 1.2. River basins between Afghanistan and Pakistan. (From: the authors)

Box 1.1. Durand Line versus international border.

Afghan authors contributing to this book follow the term 'Durand Line' while the authors from Pakistan use the term 'international border' between Afghanistan and Pakistan. This is valid for all chapters, including maps, figures and tables.

It is important to mention here that Afghanistan and Pakistan share a 2670 km boundary, which is referred to as the 'Durand Line' by Afghanistan and the 'international border' by Pakistan. The editors have given due respect to both perspectives since the book is co-produced by authors from Afghanistan and Pakistan. Both terminologies have been used by the contributors throughout this book to reflect a balanced view and to avoid any criticism which is beyond the scope of this work (Box 1.1).

Indus basin tributaries shared by Afghanistan and Pakistan have not achieved significant advances towards co-operation nor received extensive research focus. Of the nine rivers that flow across the border between Afghanistan and Pakistan, none is governed by a formal agreement or mechanism to manage shared water resources (Shroder and Ahmadzai, 2016). There is limited information available about the status of the environment, hydrology and water resources management that could be used as a starting point for dialogue on trans-boundary water co-ordination.

The strategic importance of the nine shared rivers varies

Three rivers drain into the northern province of Khyber Pakhtunkhwa in Pakistan (KPK) while the remaining six flow into the south-western province of Balochistan. On the Afghan side, all rivers originate in the Kabul province. The three rivers that cross the international border and enter KPK are larger and provide significant average annual inflows to Pakistan (>1000 million m^3 (mcm)/year) (Bhatti *et al.*, 2021). These include the Kabul, Kurram and Gomal rivers. The remaining six rivers that enter the province of Balochistan are smaller and provide relatively small average annual inflows (<100 mcm/year) (ibid.). These include the Pishin Lora, Kandar, Kand, Kadanai, Abdul Wahab Stream and Kaiser rivers. Kabul, Kurram and Gomal have the largest annual runoff in both Afghanistan and Pakistan. Likewise, infrastructure investments are greatest in Kabul, Kurram and the Gomal basin, resulting in substantially higher use of water compared to the other six transboundary rivers. Hence, the Kabul, Kurram and Gomal basins are the focus of detailed investigations in this report.

Scope and Objectives

The objective of this book is to provide a comprehensive overview of the three major Afghan–Pakistani shared river basins in terms of their water resources, land resources, ecological health, environment, climate change and the social and economic conditions for sustainable management of water resources (Fig. 1.3). This is the first book to cover major aspects of water resources management and development on the major transboundary river basins shared between Afghanistan and Pakistan, together in one volume. It is also the first book to be published after the regime change in August 2021, through which the Taliban government returned to power in Afghanistan. This power change may alter the water resources development plans of the previous (2002–2021) regime. Several chapters explore the implications of this change on water development as well as other dynamics such as the likelihood of emigration and the potential for regional co-operation. It is our sincere hope that this book reaffirms to both countries that co-operation remains an option.

Past work on these shared waters from a basin perspective has tended to focus exclusively on the Kabul basin. For example, Iqbal

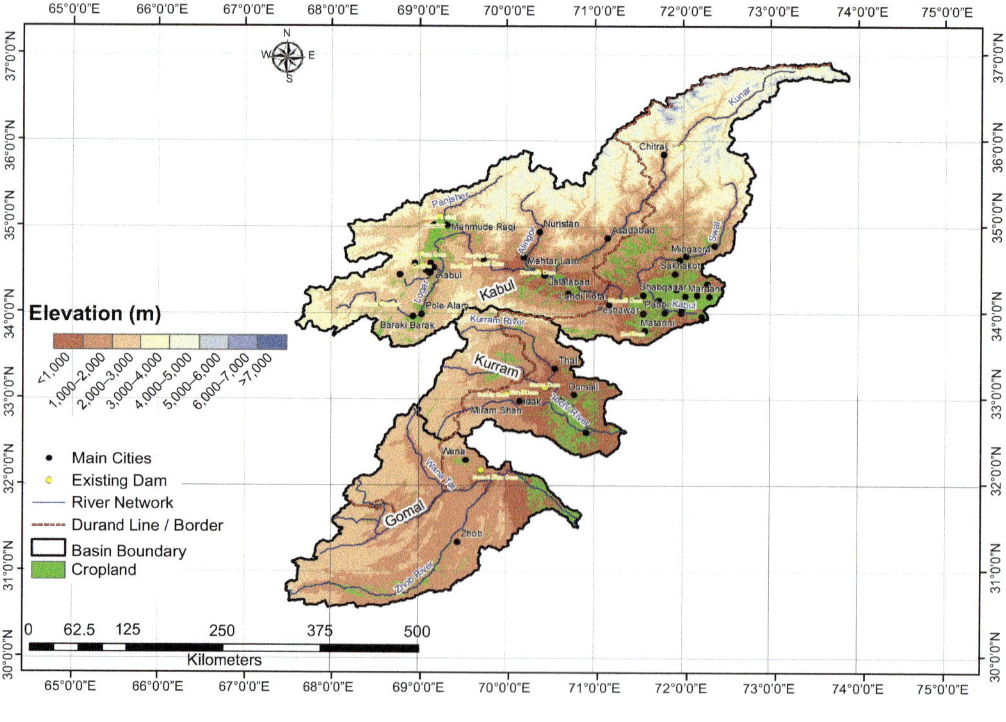

Fig. 1.3. Kabul, Kurram and Gomal river basins. (From: the authors)

(2020) talks about hydro-diplomacy in the Kabul river basin and the potential for conflict and co-operation between Afghanistan and Pakistan in this shared river basin. The research is limited to specific issues of conflict and co-operation and does not look at the broader suite of parameters related to the shared basins. Akhtar (2017) focuses on the Kabul river basin and explores the land use and land cover changes over time against the backdrop of competing uses of water and growing water demand. This research focuses on the Afghanistan portion of the Kabul river basin and does not include the Pakistan side.

This book strives to fill the gaps identified in the existing research by focusing on the three shared basins, which provide a lifeline for the people living there. The editors' and authors' aim is to make three key contributions: (i) provide a baseline assessment of the current state of these basins in terms of the socio-economic development, climate change impacts, surface and groundwater conditions, land use and land cover changes, water governance and institutional landscape; (ii) highlight the potential for co-operation for jointly developing shared water resources; and (iii) foster convergence in perspectives through the co-production process (more details in Chapter 2). Ultimately, an effort has been made to provide a balanced view of the state of the three river basins by incorporating inputs from both Afghanistan and Pakistan.

Kabul River Basin

The Kabul river basin carries special significance as it provides over 20 bcm of water each year and supports a population of some 34 million people in an area of approximately 92,000 km^2 (Masood *et al.*, 2018). The Kabul basin is the most developed among the shared rivers. Currently, it irrigates some 506,000 ha from the Kabul river on both sides of the border. This plays a critical role in local, national and regional food security.

The Kabul river basin covers around 12% of the national territory of Afghanistan and generates about 26% of the country's total annual surface water flow (Favre and Kamal, 2004). Around 9% of the total drainage area is cultivated. This plays a critical role in local, national and regional food security. Agriculture in the Kabul basin is the main contributor to food security in Afghanistan in terms of its cultivated area and irrigation potential. On the Pakistan side, the Peshawar Valley in KPK is considered the bread basket of the basin.

As a large basin with a high elevation drop, the Kabul basin has substantial potential for hydroelectricity that can support regional economic development mainly in the Afghan portion. The current installed hydropower capacity on the Kabul river basin in Afghanistan and Pakistan amounts to 465 MW (Khan and Nafees, 2018). The Kabul basin contains some of the major cities in Afghanistan including the capital city of Kabul. It is also one of the most populated river basins in Afghanistan. It covers 15 provinces and 106 districts in Afghanistan and two provinces and 21 districts in Pakistan.

The volume of water use activities just described (e.g. agriculture, hydropower, urban) are inherently interconnected as they all occur in the same basin. While it is certainly possible for the two countries to manage their portions of the basin unilaterally, it would seem to be in the interests of both countries to explore opportunities for co-ordination through a co-operative framework. At a fairly basic level, a minimum level of co-ordination could help to ensure actions on one side of the border do not undermine efforts on the other.

Kurram River Basin

The Kurram river originates in Paktia province in Afghanistan and enters Pakistan in Kurram district. The Kurram is the second most densely populated basin shared between Afghanistan and Pakistan with a total population of 5.5 million people. Over 70% of the basin area is in Pakistan with 87% of the basin population (details in Chapter 3). The population density is the highest among the three river basins. This density is above national averages, which gives it a special significance across the three basins. The water of the Kurram and its tributaries is essential for agriculture, domestic use and hydropower generation in both countries. Snowmelt from the upper part of the Kurram basin is the main source of surface flow into the river.

Of the total agricultural area in the Kurram basin, 18% is in Afghanistan and 82% in Pakistan (details in Chapter 7). The Kurram basin consists of predominantly rural settlements. The major part of the basin in Afghanistan consists of high elevations and mountainous terrain with fewer opportunities for large-scale agriculture expansion and consumptive use of water. However, on the Pakistan side, there are settlements on the plains with opportunities for further expansion of agriculture and consumptive use of water.

There is less tension on the Kurram as compared to the Kabul river basin. Pakistan has already developed dams and irrigation infrastructure. However, climate change and prolonged droughts in the past few decades may pose a threat to the viability of this infrastructure. It may be in the interests of both countries to develop mechanisms to cope with such uncertainties. Hence, this basin may provide an opportunity for engagement between the two governments.

Gomal River Basin

The Gomal river begins in the south-east of Ghazni province near Sarwandi in Afghanistan and crosses Paktia before entering the South Waziristan district of Pakistan. After leaving Afghanistan, the Gomal crosses South Waziristan district and serves as the boundary between KPK and Balochistan. The Gomal is the smallest basin in terms of geographical size. Only 1% of the population residing in this basin live in Afghanistan; the remaining 99% live in Pakistan (details in Chapter 3). A major share of the population lives in the rural areas on both sides of the border. Almost 98% of the agricultural area is in Pakistan (details in Chapter 3). There is considerable scope for investment to develop this basin; however, obstacles include continued political instability and the movement of migrants during prolonged periods of conflict in Afghanistan and security-based operations in Pakistan, particularly in South Waziristan district. Pakistan has invested in the construction of the Gomal Zam Dam in South Waziristan district which may serve as a catalyst for improving the livelihoods of people living in Dera Ismail Khan and Tank districts. Perennial canals irrigating an area of over 77,000 ha have been constructed as part of the dam command area, which could significantly improve livelihood conditions.

Like in the Kurram basin, there is little potential for large-scale agriculture development in the Gomal basin on the Afghanistan side due to the steep topography. However, there is considerable potential for hydroelectric power development, which would provide opportunities for collaboration whereby Afghanistan and Pakistan could jointly develop this infrastructure and reap the benefits of better water regulation and cheap electricity generation. Like the Kurram, there is no history of conflicts in the Gomal basin.

Conclusion

As further outlined in Chapter 3, the population of the Kabul, Kurram and Gomal river basins are significant in the national contexts of the two countries but impoverished by national and global standards. Almost one third of the population of Afghanistan resides in the three river basins and almost 15% on the Pakistan side. The *Human Development Report* 2021-22 (UNDP, 2022) ranks Afghanistan at 169 and Pakistan at 154, which indicates the poor socio-economic conditions in this region. Afghanistan (157 out of 162) and Pakistan (135 out of 162) both have relatively low gender inequality indices (UNDP, 2021). These issues are most pronounced in the Kurram and Gomal river basins where the people follow tribal norms and the majority of the population lives in rural areas.

With the fall of the Afghan Republic on 15 August 2021 and the takeover by the Taliban, conditions may change dramatically. UNDP, 2021 warned of an economic contraction because of the sudden halt in external financial support to social sectors. Under these economic circumstances, the country could be headed for humanitarian challenges (IPC 2021). Finally, there may be critical losses of capacity in Afghanistan due to emigration. As evidenced in the production of this book (see Chapter 2), the migration of Afghan technical experts may alter capacity to devise and implement development programmes that benefit the country's population.

Against this backdrop, the authors hope that this book will serve as a baseline for assessing the current state of the three basins. Each chapter highlights opportunities for collaboration between the two governments to bring people out of poverty and safeguard livelihoods. All the basins discussed in this book have the potential for sharing the benefits of joint river basin development and avoiding conflicts that may hamper prosperity. There are significant issues of data availability to perform the necessary analyses in the three basins but every effort has been made to access what data was available on both sides of the border. The authors have acknowledged limitations where data is not available.

References

Akhtar, F. (2017) Water availability and demand analysis in the Kabul River Basin, Afghanistan. Doctoral dissertation, Rheinische Friedrich-Wilhelms-Universität Bonn, Germany.

Arai, F.K., Pereira, S.B. and Gonçalves, G.G.G. (2012) Characterization of water availability in a hydrographic basin. *Engenharia Agrícola* 32(3), 591–601. DOI: 10.1590/S0100-69162012000300018.

Bhatti, M.T., Ashraf, M. and Anwar, A.A. (2021) Soil erosion and sediment load management strategies for sustainable irrigation in arid regions. *Sustainability* 13(6), 3547. DOI: 10.3390/su13063547.

Cooley, H., Christian-Smith, J., Allen, P.H. and Cohen, M. (2009) *Understanding and reducing the risks of climate change for transboundary waters*. Pacific Institute, Oakland, California and UNEP. Available at: https://pacinst.org/publication/climate-change-and-transboundary-waters/ (accessed 9 September 2022).

De Stefano, L., Duncan, J., Dinar, S., Stahl, K., Strzepek, K.M. *et al.* (2012) Climate change and the institutional resilience of international river basins. *Journal of Peace Research* 49(1), 193–209. DOI: 10.1177/0022343311427416.

Favre, R. and Kamal, G.M. (2004) *Watershed Atlas of Afghanistan*. Food and Agricultural Organization of the United National and Afghanistan Information Management Service, Kabul.

Giordano, M., Drieschova, A., Duncan, J.A., Sayama, Y., De Stefano, L, *et al.* (2014) A review of the evolution and state of transboundary freshwater treaties. *International Environmental Agreements* 14(3), 245–264. DOI: 10.1007/s10784-013-9211-8.

IPC (2021) Afghanistan: Acute Food Insecurity Situation March-May 2021 and Projection for June–November 2021. Available at: https://www.ipcinfo.org/ipc-country-analysis/details-map/en/c/1154300/

Iqbal, M. (2020) Hydro-Diplomacy in Kabul River Basin (2001-2014): The Conflict and Cooperation Potential of Water in Pakistan-Afghanistan Relations. Department of Political Science University of the Punjab, Lahore, Pakistan.

Khan, S.A. and Nafees, M. (2018) Construction of dams on Kabul River and its socio-economic implications for Khyber Pakhtunkhwa, Pakistan. *Central Asia* 83(Winter). DOI: 10.54418/ca-83.26.

Masood, A., Hashmi, M.Z. ur R. and Mushtaq, H. (2018) Spatio-Temporal Analysis of Early Twenty-First Century Areal Changes in the Kabul River Basin Cryosphere. *Earth Systems and Environment* 2(3), 563–571. DOI: 10.1007/s41748-018-0066-6.

Salman, S.M.A. (2007) The Helsinki Rules, the UN Watercourses Convention and the Berlin Rules: perspectives on international water law. *International Journal of Water Resources Development* 23(4), 625–640. DOI: 10.1080/07900620701488562.

Shroder, J.F. and Ahmadzai, S.J. (2016) *Transboundary Water Resources in Afghanistan: Climate Change and Land-Use Implications*. Elsevier Science.

Thomas, V., Azizi, M.A. and Behzad, K. (2016) *Developing Transboundary Water Resources: What Perspectives for Cooperation between Afghanistan, Iran and Pakistan?* Vol. 6. Afghanistan Research and Evaluation Unit, Kabul.

UNDP (2021) Afghanistan Socio-Economic Outlook 2021-2022: Averting a Basic Needs Crisis. Available at: https://www.undp.org/publications/afghanistan-socio-economic-outlook-2021-2022-averting-basic-needs-crisis

UNDP (2022) Human Development Report, Pakistan 2021. Available at: http://hdr.undp.org/en/countries/profiles/PAK

United Nations (2022) *Handbook on Water Allocation in a Transboundary Context*. United Nations Publications. Available at: https://books.google.com.pk/books?id=53jxzgEACAAJ (accessed July 2022).

Votrin, V. (2005) The orthodoxy and sustainable development potential for broader involvement of the orthodox churches in Ethiopia and Russia. *Environment, Development and Sustainability* 7(1), 9–21. DOI: 10.1007/s10668-003-5053-9.

Wolf, A.T., Yoffe, S.B. and Giordano, M. (2003) International waters: identifying basins at risk. *Water Policy* 5(1), 29–60. DOI: 10.2166/wp.2003.0002.

2 Methodology: Co-production of Knowledge for Ownership and Sustainability

Asadullah Meelad, Muhammad Azeem Ali Shah and Jonathan Lautze

Chapter Overview

This chapter outlines the co-production process through which this book was developed. The chapter compares its approach to other efforts to generate knowledge in transboundary basin management, and points to certain comparative advantages. Four key points that motivated the book's focus were elaborated, namely: i) new knowledge generation, ii) fostering a common perspective, iii) enhancing human development, and iv) identification of development options. The iterative process through which the book arrived at its structure, identified authors, and facilitated their joint work are also explained. Notable limitations, e.g., related to data, and challenges, e.g., political change in Afghanistan, are acknowledged, and the nuts and bolts of chapter review and evaluation are discussed. The chapter concludes with reflection on key lessons on the co-production approach applied, such as its contribution to trust-building between experts in the two countries.

Fig. 2.1. Kurram river, Mianwali district. (From: Amjad Jamal, IWMI-Pakistan)

DOI: 10.1079/9781800622371.0002

Introduction

Transdisciplinary co-production of knowledge has been a successful approach to achieving more impactful andsustainable outcomes (Clark *et al.*, 2016; Fazey *et al.*, 2018; Schneider *et al.*, 2019). Co-production is a process or methodology to jointly and collaboratively generate policy-related knowledge by engaging multiple and diverse actors (Lepenies *et al.*, 2018). The collaborative nature of this approach enables the participants to collectively address a problem and set specific goals, which enhances transformations achieved when responding to a challenge (Lang *et al.*, 2012). Ultimately, scientific knowledge generated through such an approach is likely to be more relevant and useful in policy formulation.

Unlike more traditional approaches to knowledge generation, co-production addresses the 'usability gap' through the sustained engagement of information producers and users to create credible knowledge (Cash *et al.*, 2003). The proponents of this mechanism argue that the knowledge generated through this process yields greater impact and more outcomes (Lepenies *et al.*, 2018; Norström *et al.*, 2020). However, co-production as a process can also be challenging as it brings together a wide range of actors and stakeholders from different communities, which can be more time-consuming. Equally, diverse actors will not easily embark on a collaborative effort without some degree of trust. In addition, engagement in collaborative and co-production processes often requires specific knowledge and skills to create fruitful interactions among individuals with diverse cultures, values, thoughts, experiences, skills and knowledge (Ayre *et al.*, 2018).

Sustainable water management in internationally shared river basins has always been complex, dynamic and challenging (Akamani and Wilson, 2011; Margerum and Robinson, 2015). In such contexts, knowledge created through a co-production process produces benefits well beyond the knowledge itself. By pursuing a co-production process, we can foster convergence in thinking across countries, which can help to bring key stakeholders closer to co-operation. Equally, co-production tends to be more inclusive of indigenous local knowledge, something that is often overlooked by government water management professionals in developing countries (Zarei *et al.*, 2020).

Unfortunately, co-production approaches are not always used (Maiello *et al.* 2013) and bridging the science–policy gap remains a persistent challenge. In transboundary river basins, the challenges of co-operative water management can be profound – divergent assessments on levels of water availability and use in a basin, for example, can delay progress for years. Such challenges can be partially alleviated by creating knowledge through a co-production process that helps move divergent perspectives toward common understanding (Armitage *et al.*, 2012). Under this approach, experts and stakeholders from various backgrounds and with different experiences can be engaged to explore a specific topic and produce context-related information and solutions.

A review of existing state-of-the-basin reports (SOBRs), such as those in the Nile and Mekong, drives questions on the degree to which they adhere to a co-production process. It would appear that SOBRs are often prepared by consultant teams under the auspices of their river basin organization and, while there are important consultations with national stakeholders, such interaction tends to be more oriented towards review and sign-off of new knowledge rather than direct participation in knowledge production. SOBRs published by the Nile Basin Initiative (NBI) on the Nile river basin (NBI, 2012, 2020), for example, relied on consultants contracted by NBI to draft and prepare content. Although input from country contributors was sought during the drafting process and national experts from some riparian states and the Technical Advisory Committee of the NBI validated the draft chapters (NBI, 2012), country contributors did not appear to directly draft SOBR content. Likewise, three SOBRs published by the Mekong River Commission (MRC, 2003, 2010, 2018) were mainly led and drafted by consultants to the MRC, with input, review and sign-off sought from country representatives. Ultimately, while there is clear value in securing validation and sign-off from stakeholders in a shared watercourse on a set of knowledge, more direct stakeholder generation of knowledge may enhance co-production levels.

This chapter aims to outline and explain the important steps in the co-production process through which this book was developed. This chapter first outlines the book's content and direction and how these were developed in response to the goals and purposes. Next, the authors discuss how the Afghan and Pakistani writing team was identified, assembled, commissioned and mobilized to draft specific chapters and participate in cross-chapter technical discussions. The process of chapter development is then described, in particular how data and information were collected and shared within the author team. Finally, internal and external review processes are described. The chapter concludes with a discussion of the benefits, challenges and opportunities of shared knowledge production in shared Afghan–Pakistani waters.

Rationale and Scope

This book serves four purposes. First, knowledge generation. As the output of a co-production process, it aims to generate new and credible knowledge on diverse aspects of Afghan–Pakistani shared waters. Second, a common perspective. Through an inclusive approach to assessing conditions and diagnosing problems, this book aims to promote a common perspective on the management of the water resources of the Kabul, Kurram and Gomal river basins. Third, improved human wellbeing. The creation of knowledge and a common perspective is expected to inform decision-making in Afghanistan and Pakistan, which will, it is hoped, contribute to a foundation for enhancing human development in the basins. The authors have collected and documented information summarizing the basins, including the status of their surface and groundwater resources, the impact of climate change, current developments, and social and economic circumstances. Fourth, development options. The authors look forward to outlining opportunities and pathways to enhance the human development of the basins.

To achieve its objectives, the thematic scope of this book includes a clear, complete and evidence-based account of key water and human development parameters in the basins. The reason for selecting these themes was to cover the key aspects of integrated and optimal water resource management. The scope also included gathering information and local knowledge, creating a common perspective, contributing to the improvement of human wellbeing in the basins, and highlighting the challenges and opportunities for future development.

A particular achievement of this book is the synthesis and processing of data from the two countries. As yet, there is no consolidated baseline data in either Afghanistan or Pakistan that can provide a comprehensive analysis of land, water and environment information on transboundary river basins. Data in the named areas in Pakistan is kept with public entities (e.g. Water and Power Development Authority), public universities and some individuals. Data in Afghanistan are typically kept with the Ministry of Energy and Water and the Ministry of Agriculture, Irrigation and Livestock, public and private universities, and individual experts.

This book will be revised and updated through follow-up state-of-the-basin books in the future. In the meantime, this state-of-the-basin book is expected to bolster scientific and evidence-based understanding of the basins and create a dialogue among stakeholders who will determine the future of Afghan–Pakistani river basins. In these chapters, the authors cover the following three river basins in their entirety:

- Kabul river basin
- Kurram river basin
- Gomal river basin

Book Structure and Development

The development of this book is supported by the Water Management for Enhanced Productivity project (WMfEP), which commenced in July 2018 and is scheduled to be completed in June 2023. The project is funded by the US Agency for International Development's Mission to Pakistan (USAID/Pakistan) and implemented by the International Water Management Institute (IWMI) in partnership with the government of Khyber Pakhtunkhwa (KPK). The WMfEP's activities were carried out to reduce constraints to the productive and sustainable use of water for agricultural production in KPK. Improved arrangements for water governance

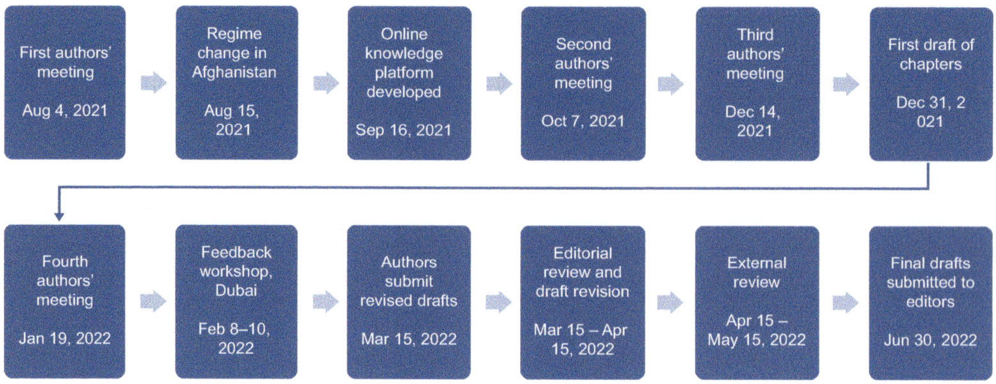

Fig. 2.2. The book production workflow.

and management will enable increased farm household income and improved livelihoods to contribute to socio-economic development and political stability in the KPK and beyond (). The geographical scope of the WMfEP covered the entire KPK including merged districts. All these areas are located in the three river basins covered in this book. Achieving the project's goals would not have been possible without a collaborative, integrated approach to issues of water management. Therefore, Component 4 of the WMfEP focused on increasing cross-border collaboration on water policy, practices and water sector challenges common to each country. The objective of Component 4 was to increase opportunities for sector professionals in each country to develop skills, share experiences and knowledge, and increase understanding of water policies, water sector incentives and approaches to drive the adoption of water-efficient agricultural practices, technologies, watershed management, improved water resources management, and increased dialogue on transboundary water benefit-sharing. The book production workflow and timeline is presented in the Fig. 2.2.

Reviewing international experience to inform content structure

Before developing the outline for this book, the editorial team reviewed existing SOBRs for other transboundary river basins. This review included eight SOBRs from four river basins: Delaware (Delaware River Basin Commission 2008), Mekong (MRC, 2003, 2010, 2018), Nile (NBI, 2012, 2020) and Saskatchewan (Partners for the Saskatchewan River Basin, 2009). A review of other SOBRs was used to: (i) guide the structure of the present SOBR on Afghanistan–Pakistan river basins, manifested in the table of contents; and (ii) understand the methodology typically applied in the development of thematic chapters. The review also helped the editorial team understand the required information on the geography, hydrology, flora and fauna of the basin and the social and economic conditions of its peoples. The review indicated the following main points:

1. SOBRs aim to provide a general view of a river basin including its hydrological, climatic, geographic and socio-economic circumstances, and the status of existing co-operation or the potential for future co-operation between and among the riparian states.
2. SOBRs provide an assessment of the conditions related to the development, management and use of water and other natural resources within a basin and its related impacts.
3. SOBRs are also intended to provide a list of suggested actions for the optimal and sustainable development of an entire transboundary river basin.
4. The main aims of developing an SOBR are: (i) evaluation of the adequacy and effectiveness of existing management strategies and practices; and (ii) stimulating policy discussions

on alternative management strategies and practices.

5. SOBRs are flagship products, which can be produced every several years based on the facts and figures from the available data and information for a transboundary river basin.

To build on the existing experiences of basin assessment, the editorial team also conducted a literature review of existing and publicly available transboundary diagnostic analysis (TDA) reports, which were mostly focused on African river basins. The TDAs reviewed included those for the Tuli Karoo basin shared between Botswana, South Africa and Zimbabwe (Ebrahim *et al.*, 2019) and the Shire river basin shared between Malawi and Mozambique (Chairuca *et al.*, 2019). This review of TDAs indicated the following points.

- TDA reports help riparian states develop a common understanding of the water resources and their issues in a given transboundary river basin, which will lead to co-operative management in the basin.
- Existing TDAs are focused on shared surface and groundwater resources and their related socio-economic challenges.
- TDAs highlight the main issues a transboundary river basin faces and help create the necessary conditions for joint action by basin states.

Ultimately, structures used to organize the content of TDAs, basin books and other SOBRs were reviewed to determine and sequence the topics to be discussed. Tables of contents in the reviewed TDAs indicated the adoption of a mostly similar structure, which included socio-economics, climate, water resources (both surface and ground), land and water uses, institutions and governance. Therefore, this state-of-basin report largely adopted structures commonly used in already published SOBRs and TDAs. Based on this extensive literature review, this book is organized into 10 chapters with the preliminary titles shown below:

1. Introduction
2. Methodology
3. Demography and Socio-economics
4. Climate
5. Surface Water
6. Groundwater
7. Land and Water Use
8. Institutions and Governance
9. Co-operation in the Afghan–Pakistani River Basins
10. Conclusion and Recommendations

Chapter 1 summarizes the broader international context of the three transboundary river basins and the significance of Afghanistan–Pakistan shared waters. It also provides the rationale for choosing the three basins and summarizes all the chapters.

Chapter 2 explains the co-production methodology employed for the development of this book. This is the first book written on the subject of Afghanistan–Pakistan shared waters and was co-developed by experts from both sides of the basin. The process was designed so that all the existing knowledge on the three shared basins is synthesized to provide a baseline for further work on these basins. The chapter explains the complete co-production process of data collection, tools and techniques employed in the writing-up.

Chapter 3 discusses the level of socio-economic development, challenges and opportunities that exist in the three basins. Key parameters that receive focus include demography, poverty, health, food security, livelihood, employment, economic growth, energy security, human development and overall economic conditions.

Chapter 4 discusses the state of the climate in the three river basins. There are huge data gaps stemming from the long history of conflict in Afghanistan. These limitations are discussed to carry out a meaningful modelling analysis for future climate projections. The authors discuss climate variability and trends in the past few decades and, based on historical data, provide insights into future climate change projections. They also discuss climate governance in the basin states and implications for future development.

Chapter 5 discusses surface water resources. The chapter is organized into three sections, each dealing with the Kabul, Kurram and Gomal basins. Within each section, the authors discuss the hydrology of the basin based on actual flow data. The authors also discuss changes in stream flow over time and likely future scenarios as well as water-quality issues on both sides of the basins. The chapter also describes the existing reservoirs and their locations in each basin and the surface water and groundwater interactions and implications for the future.

Chapter 6 provides an overview of the groundwater situation. The authors begin with a discussion on groundwater aquifers and the state of the water table in basin states. They then look at groundwater variation over time and changes in groundwater quality across the region. The authors then turn their attention to groundwater use on both sides of the border and the dependence of various economic sectors on groundwater. They conclude with a discussion of groundwater management challenges with key messages for future considerations.

Chapter 7 provides a detailed analysis of land and water use in neighbouring countries. There are maps on land use, land cover and cultivated areas on both sides of the border. The authors discuss agricultural and non-agricultural uses of water and present possible development pathways.

Chapter 8 provides a detailed analysis of legal and institutional structures. Both countries have distinct legal structures under which water institutions function. Water governance in Afghanistan consists of a complex mix of formal and informal laws and institutions. Pakistan is a federation and water is a provincial matter. The authors map institutions in both countries according to the functions they perform and the commonalities and differences in water governance. Some legal architecture has recently evolved with new laws and institutions put in place and the authors analyse how these might be used for better transboundary water governance.

Chapter 9 presents a picture of the co-operation between Afghanistan and Pakistan on shared river basins. The authors begin with the current status of (non-)co-operation between the two states in managing their shared water resources and discuss the consequences of non-co-operation and the potential barriers to co-operation on shared water resources. They then present some of the opportunities and benefits of co-operation along with guiding principles and international frameworks for co-operation on shared river basins. The key message is that Afghanistan and Pakistan have more to gain by sharing the benefits of co-operation rather than dividing these resources as Pakistan did with India, with the Indus Water Treaty.

Chapter 10 summarizes the main messages from each chapter for policymakers willing to work on developing a co-operative framework for water sharing between Afghanistan and Pakistan and links this co-operative framework with the UN Sustainable Development Goals, in particular Goal 6: 'Ensure availability and sustainable management of water and sanitation for all'.

Identification of Experts and Co-authorship

In each chapter, every effort was made to form a team of at least three experts; a leading expert from Afghanistan and Pakistan, and an international expert from IWMI. The aim was to create technical dialogue and discussions between water professionals from the two countries as they developed the chapters assigned to them. Each expert was expected to provide all the data he/she had access to for enriching that particular chapter.

The contributing experts from Afghanistan and Pakistan were carefully selected from the research, academic and practitioner communities to participate with the international experts from IWMI in writing up the chapters. A long list of potential contributing authors was narrowed to a list of those qualified, available and willing to participate in the process. Each expert was selected on the basis of his/her education, experience, skills in water resources management and familiarity with the river basins.

Unfortunately, sustained participation of Afghan experts faced at least two challenges. First, the regime change of 15 August 2021 – which occurred in the early stages of chapter development – disrupted the lives of certain co-authors and caused them to refocus their time and energy on essential issues such as immigration. Second, certain potential authors expressed concerns about engaging on the topic due to the historically troubled relations between the two concerned countries – both before and after the regime change. Despite these

Table 2.1. Chapter authors.

Chapter	Author	Country
1	Azeem Shah Jonathan Lautze Asadullah Meelad	Pakistan International Afghanistan
2	Asadullah Meelad Azeem Shah Jonathan Lautze	Afghanistan Pakistan International
3	(Ms) Marie-Charlotte Buisson Alias Wardak Inayat ullah Jan	International Afghanistan Pakistan
4	Zia Hashmi Mujib Ahmad Azizi Tousif Bhatti	Pakistan Afghanistan Pakistan
5	Arif Anwar Hassaan Furqan Khan	Pakistan
6	Shahid Iqbal Paul Pavelic	Pakistan International
7	Taimoor Akhtar Karthikeyan Matheswaran	Pakistan International
8	(Ms) Fazilda Nabeel Azeem Shah	Pakistan
9	Jonathan Lautze Shakeel Hayat Asadullah Meelad	International Pakistan Afghanistan
10	Jonathan Lautze Azeem Shah Asadullah Meelad	International Pakistan Afghanistan

challenges, the majority of chapters contain an Afghan co-author (Table 2.1).

Knowledge Platform and Data-processing

During the first authors' meeting on 4 August 2021, it was decided that IWMI would create a knowledge platform for the shared river basins of Afghanistan and Pakistan to support chapter development. The initial version was made available to the contributing authors on 16 September 2021. This platform contains detailed resources, especially remote sensing/GIS and hydrological data. The authors were able to access the data and add and share their data with others.

The data used in the analysis in Chapter 3 has been extracted from the National Statistic and Information Authority for Afghanistan and the Pakistan Bureau of Statistics. These two sources are the most reliable sources for this section. However, there were some other sources used such as the Afghanistan District Dashboard developed by the World Bank Group and the Pakistan Social and Living Standards Measurement Survey. For the discussion on migration, UNHCR, UNDESA and International Organization for Migration data have been used as their data is often used in academic works. The discussion on socio-economic conditions of the three basins relies on data from the UNDP, World Bank Group, Integrated Food Security Phase Classification, Global Development Indicators, UNICEF and some formal sources from Pakistan and Afghanistan. Even though this data was not tailored for the three basins, it provides relevant facts and information on socio-economic conditions. The authors adapted this data using their own calculations and analysis.

For Chapter 4, besides the published literature, the data from the Department of Metrology and Hydrology of Afghanistan, UNEP, High Asia Refined Analysis and published literature was used. The authors also used data from 21 meteorological stations in Afghanistan and 10 stations in Pakistan. Unfortunately, as the analysis of climate requires data over long periods, there were limited data for Afghanistan due to the long history of conflict.

For Chapter 5, data from multiple sources were used such as formal sources (e.g. Pakistan Water and Power Development Authority), and figures and statistics from published articles and online datasets. Chapter 6 mostly used the data and information from published scientific hydrogeological studies. Chapter 7 relies mainly on data derived from the following two remote sensing datasets:

i. SERVIR-HKH Regional Cover Monitoring System: this dataset has a 30 m spatial resolution, which is used to estimate land use and land cover.
ii. Synthesized Global E: this dataset has a 1 km spatial resolution, which is used for estimating agricultural water use.

In addition, secondary data was collected from multiple published sources.

The discussion in Chapter 8 on water governance and its institutions relies on, to a large extent, legal and policy documents, including the fundamental, federal, national and subnational legal statutes. The data needed for this chapter was collected from official and publicly available sources. These statutes and policy documents regulate the tasks and responsibilities of water institutions in Afghanistan and Pakistan and describe the rules of governance, decision-making mechanisms, and the overall management methods of water resources in the three basins. The data and information used in Chapter 9 were collected from various publicly available academic sources.

Data limitations

While this is the first book of its kind written in cross-border co-production on the Kabul, Kurram and Gomal river basins, it may possess limitations. Limited data and information were available on all three catchments, but particularly the Kurram and Gomal. As also acknowledged by the authors in Chapter 4, long-term, reliable and adequate climate-related data and information are still lacking in these river basins. The climate-related monitoring network has yet to be adequately expanded, especially in the uppermost parts of the basins where the snowfall happens and glaciers exist. The authors stated that even though climate observation and data collection was begun in the middle of the 20th century, the prolonged conflicts disrupted the process and we now face scattered and insufficient data.

Similarly, the authors in Chapter 5 pointed out the limited availability of hydrologic datasets and scientific studies for the Kurram and Gomal river basins. As a solution, they have relied more on grey literature in these two river basins and most of their analysis is focused on the Kabul basin, which is in a much better situation interim of data availability compared to the Kurram and Gomal basins. Likewise, the authors of Chapter 6 faced a serious scarcity of data regarding the existence, size and quality of groundwater aquifers and zones across the three basins. Both the quantity and quality of groundwater resources are not adequately monitored. Nonetheless, it is hoped that this book will provide a foundation for future work and experts will continue to build on the insights offered. Equally, in Chapter 7, the authors acknowledge the insufficiency of relevant water- and land-use data; there were some provincial- and national-level land-use data but it was not useful for basin-level analysis.

Chapter Development

IWMI convened regular meetings of authors. On 4 August 2021, IWMI hosted the first virtual meeting of contributing authors. The discussions focused on international experience with state-of-basin reports, a brief overview of the Afghanistan–Pakistan data availability, data-sharing, and timelines for completing the report.

After the first meeting, the authors were asked to discuss the initial draft outline developed from the literature review and offer their revised outlines for each chapter by the end of September 2021. During this process, each group of authors had internal discussions on the draft outline and submitted their final feedback to the IWMI team.

After receiving the authors' feedback, the IWMI team organized the second meeting on 7 October 2021. The discussions in this meeting focused on chapter outlines, a template and the creation of a shared drive for data-sharing among the authors.

According to the agreed timeline, the first draft of the chapters was planned to be finished in December 2021. To follow up on this, IWMI organized a virtual meeting on 14 December 2021. There was another virtual meeting in January 2022 to check on progress and make preparations for a workshop in February 2022.

IWMI's team organized and hosted a feedback workshop from 8–10 February 2022 in Dubai with in-person and virtual attendance (Fig. 2.3). This workshop was intended to bring together all the contributing authors and get feedback and comments from each author, the editors and other experts.

During this workshop, each group of authors delivered a detailed presentation on their chapters to stimulate discussion and to get

Fig. 2.3. Author group photograph in Dubai.

Fig. 2.4. Authors' working group meeting.

feedback from the participants (Fig. 2.4). All the comments and feedback were recorded by IWMI and shared with the authors after the workshop.

Once the authors incorporated feedback from the workshop, the editorial team reviewed the revised chapters. Further comments on all the chapters were provided to all the authors and a deadline was given to revise the drafts. Once the revised drafts were received, the editors went through another round of review and prepared versions for external reviewers. These draft chapters were then submitted to external reviewers for an independent expert review.

Following external reviews, the authors revised their chapters and submitted them to the IWMI editorial team. The editorial team did two final rounds of revision and prepared all chapters for submission to the publisher.

Conclusions

This book is the outcome of a long and at times challenging journey of co-production. This is the first time such a process has been undertaken in the shared river basins of Kabul, Kurram and Gomal. Despite the challenges, the process mobilized and engaged a diverse group of experts from both Afghanistan and Pakistan in a collaborative initiative. It comprised a significant experience in which water experts from both countries were able to put aside the historical mistrust between the two states to assess the status of water resources and lay the basis for future co-operation. The process was effective in the following ways:

- The co-production approach was successfully applied. Experts preparing future state-of-basin reports and transboundary diagnostic analysis now have an additional template for a co-production mechanism to address the usability gap of knowledge in shared river basins.
- Participants learned that trust-building is the core component of a successful co-production process. The process created a significant degree of confidence, trust and good working relations among the contributing authors. Such trust can play an important role in laying the foundation for future collaborations.
- The process created solid, salient and reliable knowledge and information that can function as a baseline for better scientific understanding of the basins for similar works in the future.
- This work identified key issues and challenges within the shared basins using a replicable process.

As is the case with many ambitious endeavours, there were challenges during the course of this journey, which affected both the process and its outcomes. The regime change in Afghanistan took place during the process of this co-development. This situation directly and negatively affected the participation of Afghans. For instance, Afghan-based authors were not able to participate in person in the Dubai workshop due to a visa ban and the pending international recognition of the new regime in Kabul. Despite this, the co-production process progressed and concluded productively. It has created a clear and credible baseline for further knowledge generation and collaborations in the future.

References

Akamani, K. and Wilson, P.I. (2011) Toward the adaptive governance of transboundary water resources. *Conservation Letters* 4(6), 409–416. DOI: 10.1111/j.1755-263X.2011.00188.x.

Armitage, D., de Loë, R. and Plummer, R. (2012) Environmental governance and its implications for conservation practice. *Conservation Letters* 5(4), 245–255. DOI: 10.1111/j.1755-263X.2012.00238.x.

Ayre, M.L., Wallis, P.J. and Daniell, K.A. (2018) Learning from collaborative research on sustainably managing fresh water: implications for ethical research–practice engagement. *Ecology and Society* 23(1). DOI: 10.5751/ES-09822-230106.

Cash, D.W., Clark, W.C., Alcock, F., Dickson, N.M., Eckley, N. *et al.* (2003) Knowledge systems for sustainable development. *Proceedings of the National Academy of Sciences of the United States of America* 100(14), 8086–8091. DOI: 10.1073/pnas.1231332100.

Chairuca, L., Chintengo, P., Ebrahim, G., Fraser, C., Lautze, J. *et al.* (2019) *Transboundary Diagnostic Analysis of the Shire River Aquifer System*. Groundwater Management Institute and International Water Management Institute. Available at: https://sadc-gmi.org/wp-content/uploads/2020/05/ENG_ShireConWat-TDA-1-3.pdf (accessed 10 July 2022).

Clark, W.C., Van Kerkhoff, L., Lebel, L. and Gallopin, G.C. (2016) Crafting usable knowledge for sustainable development. *Proceedings of the National Academy of Sciences* 113(17), 4570–4578. DOI: 10.1073/pnas.1601266113.

Delaware River Basin Commission (2008) *Delaware River State of the Basin Report*. Delaware River Basin Commission. West Trenton, NJ, pp. 1–85.

Ebrahim, G., Father, F., Hlatywayo, C., Keetile, K., Kenabatho, P. *et al.* (2019) *Transboundary Diagnostic Analysis (Baseline Report) for Tuli Karoo System*. USAID, International Water Management Institute, Groundwater Management Institute and LIMCOM. Available at: http://conjunctivecooperation.iwmi.org/wp-content/uploads/sites/38/2021/06/TuliKarooTDA-compressed.pdf (accessed 10 July 2022).

Fazey, I., Schäpke, N., Caniglia, G., Patterson, J., Hultman, J. *et al.* (2018) Ten essentials for action-oriented and second order energy transitions, transformations and climate change research. online, *Energy Research & Social Science* 40, 54–70. DOI: 10.1016/j.erss.2017.11.026.

IWMI (2020) "*Water Management for Enhanced Productivity, Task 4.2.3: Afg-Pak River basins-State of the Basins Report*" (Task Report). Submitted to USAID Pakistan on 1 December 2020.

Lang, D.J., Wiek, A., Bergmann, M., Stauffacher, M., Martens, P. *et al.* (2012) Transdisciplinary research in sustainability science: practice, principles, and challenges. *Sustainability Science* 7(S1), 25–43. DOI: 10.1007/s11625-011-0149-x.

Lepenies, R., Hüesker, F., Beck, S. and Brugnach, M. (2018) Discovering the political implications of coproduction in water governance. *Water* 10(10), 1475. DOI: 10.3390/w10101475.

Maiello, A., Viegas, C.V., Frey, M. and Ribeiro, J.L.D. (2013) Public managers as catalysts of knowledge co-production? Investigating knowledge dynamics in local environmental policy. *Environmental Science & Policy* 27, 141–150.

Margerum, R.D. and Robinson, C.J. (2015) Collaborative partnerships and the challenges for sustainable water management. *Current Opinion in Environmental Sustainability* 12, 53–58. DOI: 10.1016/j.cosust.2014.09.003.

MRC (2003) *State of the Basin Report 2003 (Executive Summary)*. Mekong River Commission, Phnom Penh. Available at: https://doi.org/10.52107/mrc.akbo6q (accessed September 2022).

MRC (2010) *Annual Report*. Mekong River Commission. Vientiane, Laos. Available at: https://www.mrcmekong.org/assets/Publications/governance/Annual-Report-2010.pdf (accessed July 2022).

MRC (2018) *State of Basin Report*. Mekong River Commission. Vientiane, Laos. Available at: https://www.mrcmekong.org/assets/Publications/governance/Annual-Report-2010.pdf (accessed July 2022).

NBI (2012) *State of the River Nile Basin 2012*. Nile Basin Initiative, Entebbe, Uganda, pp. 1–134.

NBI (2020) *State of the River Nile Basin 2020*. Synthesis. Nile Basin Initiative, Entebbe, Uganda.

Norström, A.V., Cvitanovic, C., Löf, M.F., West, S., Wyborn, C. *et al.* (2020) Principles for knowledge co-production in sustainability research. *Nature Sustainability* 3(3), 182–190. DOI: 10.1038/s41893-019-0448-2.

Partners for the Saskatchewan River Basin (2009) *From the Mountain to the Sea: The State of the Saskatchewan River Basin*. Saskatoon, Canada. Available at: https://www.saskriverbasin.ca/wp-content/uploads/2021/06/State_of_the_Basin_Report.pdf (accessed 10 July 2022).

Schneider, F., Giger, M., Harari, N., Moser, S., Oberlack, C. *et al.* (2019) Transdisciplinary co-production of knowledge and sustainability transformations: Three generic mechanisms of impact generation. *Environmental Science & Policy* 102, 26–35. DOI: 10.1016/j.envsci.2019.08.017.

Zarei, Z., Karami, E. and Keshavarz, M. (2020) Co-production of knowledge and adaptation to water scarcity in developing countries. *Journal of Environmental Management* 262, 110283. DOI: 10.1016/j.jenvman.2020.110283.

3 Demography, Human Development and Economic Conditions

Marie-Charlotte Buisson, Alias Wardak and Inayatullah Jan

Chapter Overview

The Kabul, Kurram, and Gomal river basins are home to approximately 43 million (authors' own calculation) people who experience a diverse set of human and economic development realities. This chapter draws profiles of the basins in terms of demography and human and economic development. In doing so, the authors illustrate the human realities that overlie the hydrological units. This chapter presents basin-level data by combining and rescaling data for specific administrative units in Afghanistan and Pakistan. The authors find disparities between and within the two countries in terms of population size, migration dynamics, economic indicators and production of power. Despite these differences, people share similar challenges in achieving food security, poverty alleviation, and mitigation and adaptation to climate change. Improved, co-ordinated strategies and policies have the potential to strengthen the governance of these transboundary river basins and deliver better outcomes for people and resource sustainability.

Fig. 3.1. Women's gathering Gomal Zam Area, Pakistan. (From: Khadija Begum, IWMI-Pakistan)

DOI: 10.1079/9781800622371.0003

Introduction

The Kabul, Kurram and Gomal river basins are home to approximately 43 million inhabitants who share multiple resources that play a central role in development and economic life. People living in these basins experience some similarities in terms of demography and religion. Nonetheless, there is considerable disparity in development outcomes between Afghan and Pakistani portions despite the fact that the two countries ultimately rely on the same resources. Consistency and variation in demography and human and economic conditions across the two countries sharing the basins have not been extensively investigated, yet accounting for these realities is critical as these conditions reflect the ultimate outcomes to which water management contributes.

This chapter characterizes the demography, and human and economic development in the three basins. The objective is to describe the human realities in the hydrological units. The chapter is based on combining and rescaling secondary data from the two countries. While existing data are collected for specific administrative units in Afghanistan and Pakistan, data are presented in a basin frame to support a basin-level vision and analysis.

Province- and district-level secondary data used were taken from the National Statistics and Information Authority (NSIA) for Afghanistan and from the Pakistan Bureau of Statistics (PBS). In addition, the authors have used data and information published in the grey literature during the last few years for the two countries.

The second section describes the demography and population dynamics of the river basins, including population density and migration trends. The third section analyses socio-economic and poverty indicators and the national economic contexts, growth and employment. Section 4 zooms into the key economic sectors – agriculture, industry and power. The conclusion summarizes the results and proposes ways forward.

Population and Demographics

Population structure

The Kabul river basin is the largest of the three transboundary basins and the most populated. Covering 15 provinces and 106 districts in Afghanistan and two provinces, and 21 districts of Pakistan, the basin hosts 33,994,000 inhabitants on both sides of the border (Table 3.1). In comparison, the Kurram and Gomal basins are smaller, with populations of 5,558,000 and 3,845,000 inhabitants, respectively.

The largest share of area and population are in Pakistan for all three basins (Fig. 3.2). For the Gomal basin, 8% of the area is in Afghanistan with only 1% of the basin's total population. In the case of the Kabul basin, while the area is almost equally split between the two sides of the border (53% in Pakistan and 47% in Afghanistan), the Pakistani (downstream) portion of the basin is more populated than the Afghan side with 72% of the total Kabul basin population. The same trend is observed for the Kurram basin with 29% of the basin area in Afghanistan and 13% of the total basin population.

While the basins cover roughly similar portions of the two countries, the proportion of the population reliant on these basins is relatively greater in Afghanistan. In Afghanistan and Pakistan, the aggregated area of the three basins

Table 3.1. Area and population of the Gomal, Kabul and Kurram river basins. (Based on National Statistics and Information Authority data (NSIA, 2021) for Afghanistan; Pakistan Bureau of Statistics census data (PBS, 2017) for Pakistan)

	Area (km^2)*			Population (millions)		
	Afghanistan	Pakistan	Total	Afghanistan	Pakistan	Total
Gomal basin	2,300	2,610	28,300	49,000	3,796,100	3,845,200
Kabul basin	86,000	96,200	182,100	9,557,000	24,437,000	33,994,000
Kurram basin	5,900	14,700	20,600	748,100	4,809,600	5,557,700

*Area calculated on the basis of administrative units and may differ from the watershed area.

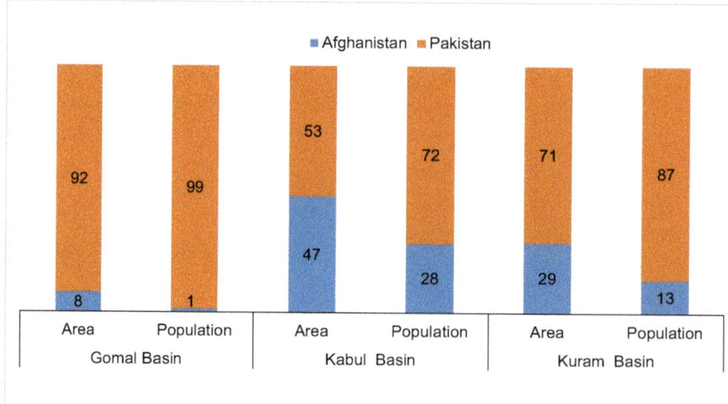

Fig. 3.2. Percentage of Kabul, Kurram and Gomal basin area and population by country. Area calculated on the basis of administrative units which may differ from watershed area. (Based on National Statistics and Information Authority data (NSIA, 2021) for Afghanistan; Pakistan Bureau of Statistics census data (PBS, 2017) for Pakistan)

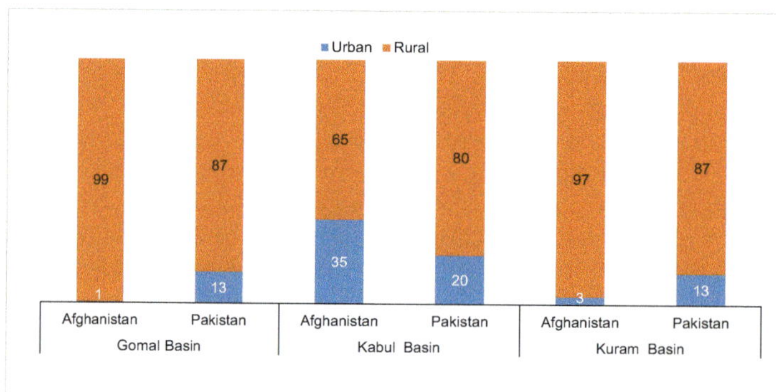

Fig. 3.3. Percentage of urban and rural population in Kabul, Kurram and Gomal river basins by country. (Based on National Statistics and Information Authority data (NSIA, 2021) for Afghanistan; Pakistan Bureau of Statistics census data (PBS, 2017) for Pakistan)

occupies about 14% and 16% of the territory. In terms of population, amongst the approximately 36 million inhabitants of Afghanistan, 29% live in one of the river basins. In comparison, in Pakistan, only 1% of the approximately 221 million Pakistanis live in one of the three basins.

The average annual population growth rate is greater in the Pakistani portions of the basins than the national average. While the population of Pakistan grew annually at an average rate of 2.8% between 1998 and 2017, the rate was 3.7% within the Kabul, Kurram and Gomal basin districts. In Afghanistan, figures for population growth were not available at the provincial or district level; however, the national figure shows a higher growth rate than that of Pakistan over the same period, with an average annual rate of 4.4%.

The three basins are largely rural and only a small share of the population lives in urban areas (Fig. 3.3). For Kurram and Gomal basins, for example, 12% of the population live in urban areas while 88% are in rural settings. The urban share of the basin population is slightly higher for the Kabul basin (Kabul and Peshawar drive the trend upward). In the Kabul basin, 24% of the population lives in cities: 35% and 20% in Afghanistan and Pakistan. Excluding Kabul province, river basin districts in Afghanistan are more rural than those in Pakistan. Finally, the population from the

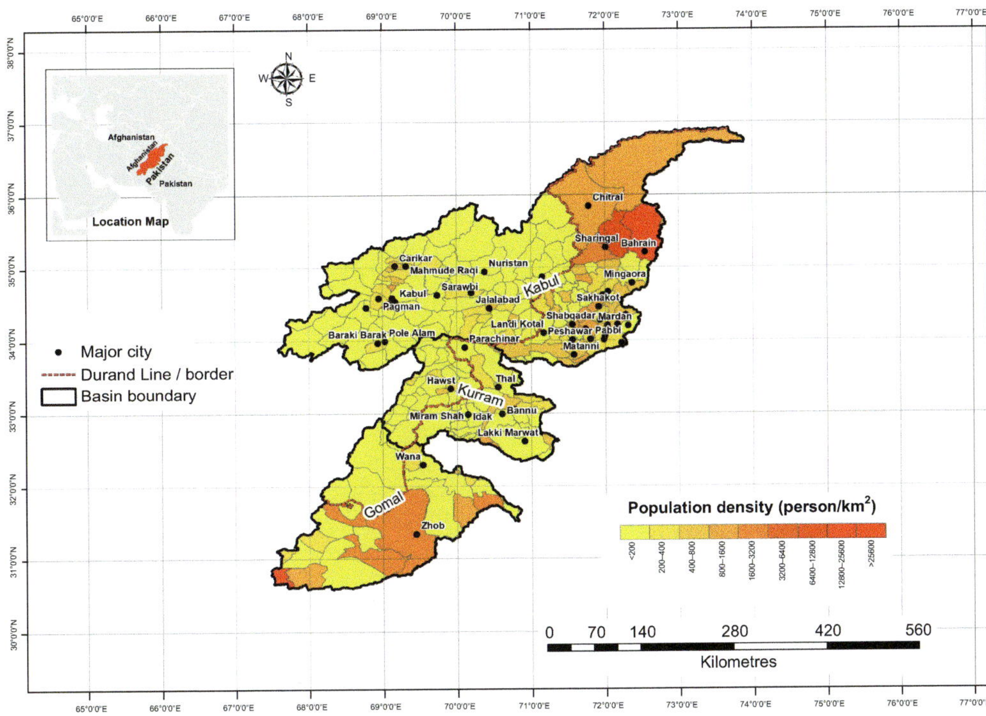

Fig. 3.4. Population density of Kabul, Kurram and Gomal river basins. (Based on National Statistics and Information Authority data (NSIA, 2021) for Afghanistan; Pakistan Bureau of Statistics census data (PBS, 2017) for Pakistan)

three basins in both Afghanistan and Pakistan shares the same ethnolinguistic characteristics as most of the population is Pashtun.

Population density

The Kurram basin is the most densely populated. Population density is 186 per km^2 in the Kabul basin, 270 per km^2 in the Kurram basin and 106 per km^2 in the Gomal basin. The density is, overall, lower in Afghanistan. While the three basins are mostly rural, there is still a high range in the population density between districts. In Afghanistan, there are almost 10,000 inhabitants per km^2 in Kabul City district. Remote districts such as Badakhshan and Nuristan have a population density below 10. The gap between maximum and minimum values is slightly lower for Pakistan but ranges from 6.4 inhabitants per km^2 in the northern areas to 3395 in Peshawar district (Fig. 3.4).

The three basins tend to be slightly less densely populated in Pakistan than the rest of the country, while the reverse is the case in Afghanistan. In Afghanistan, the national density is 60 per km^2 as compared to 22 for the Gomal basin, 111 for the Kabul basin, and 127 for the Kurram basin. In Pakistan, there are, on average, 287 inhabitants per km^2 but in the Pakistani districts of the Gomal basin, there are 146 inhabitants per km^2, 254 in the Kabul basin, and 327 in the Kurram basin.

Migration Dynamics

The history of migration in Afghanistan goes back to the end of the 1970s when the Soviet Union invasion forced millions of Afghans to leave

Table 3.2. Migration from Kabul, Kurram and Gomal river basins in Pakistan. (Based on National Statistics and Information Authority data (NSIA, 2021) for Afghanistan; Pakistan Bureau of Statistics census data (PBS, 2017) for Pakistan)

Basin	Rural	Urban	Total
Kabul	626,900 (84%)	119,800 (16%)	746,800
Kurram	109,900 (89%)	13,900 (11%)	123,800
Gomal	57,500 (93%)	4,600 (7%)	62,100
Total	794,300 (85%)	138,300 (15%)	932,700

their homes and seek refuge in foreign countries, mainly Pakistan and Iran, where up to five million Afghans lived during the 1980s and 1990s (IFO Institute, 2021). From 2002 onwards, around 5.3 million Afghans returned home but continued conflict and the rise of poverty have resulted in a decline in the number of returnees in recent years (UNHCR, 2022) with only 1363 people in 2021. Besides political and economic factors, there are also demographic and ecological dynamics that force Afghans to leave their country (IFO Institute, 2021).

According to UNHCR (2022), there are 2.6 million Afghans registered as refugees worldwide, of which 2.2 million are living in Pakistan and Iran and 3.5 million are internally displaced people. The intensified conflict in 2021 leading to the government's collapse in August 2021 coupled with ongoing drought and economic constraints displaced over 600,000 people from January to September 2021 who are in urgent need of humanitarian assistance (IPC, 2021a). Since January 2021, 128,800 Afghans arrived in neighbouring countries (61% in Pakistan, 25% in Iran and the remaining sought refuge in Tajikistan and Uzbekistan) (UNHCR, 2022). Most newly arrived refugees in Pakistan in 2021 are from the transboundary river basins, especially the provinces of Nangarhar (29%), Kabul (21%), Kunduz (8%), Ghazni (6%) and Laghman (5%). These refugees cited violence and conflict as the main reasons for their migration but 18% also mentioned the lack of livelihood opportunities and another 15% the lack of land.

Since 1990, Pakistan has hosted one of the five largest refugee populations worldwide (IOM, 2018). At the same time, labour emigration constitutes an important aspect of mobility in Pakistan (Cibea *et al.*, 2013). The nature of migration from Pakistan differs fundamentally from that in Afghanistan. While migration from Afghanistan was mostly for security reasons, migration from Pakistan is mostly driven by economic concerns such as employment abroad or in other parts of the country or because of risks and shocks due to climate change. According to the Labour Migration Report of the Government of Pakistan (GoP, 2020), the main destinations for Pakistani migrants are Gulf countries and Europe. Until 2020, 6.13% of the labour force migrated from Pakistan to the Gulf Cooperation Council countries including Saudi Arabia (61%) and UAE (24%). The remaining labour force migrated to European countries, Malaysia, the USA and Africa (GoP, 2020).

Zooming into the Kabul, Kurram and Gomal basins highlights a clear distinction between emigration from rural and urban areas in Pakistan. Of the approximately 746,800 emigrants from the Kabul basin in 2017, 84% were from rural areas. A similar picture can be seen in the Gomal and Kurram basins (Table 3.2). These differences highlight the economic conditions of rural areas within the basins where unskilled labourers have limited prospects of employment in local markets and prefer to go abroad to support their families (Hussain and Jan, 2020).

Climate change-induced migration, particularly in relation to droughts and floods, has become a common phenomenon in recent years forcing Afghans and Pakistanis to leave their homes and seek refuge inside and outside the country (Ishfaq *et al.*, 2017). With projected temperature increases, changes in timing and volume of precipitation, and intensification of climate hazards, the number of migrants is expected to rise.

Table 3.3. Development indicators of Pakistan and Afghanistan (Adapted from UNDP, 2020)

Country/region	HDI value	HDI rank	Life expectancy at birth	Expected years of schooling	Mean years of schooling	GNI per capita (2017 PPP US$)
Afghanistan	0.511	169	64.8	10.2	3.9	2.2
Pakistan	0.557	154	67.3	8.3	5.2	5.0
South Asia	0.641	n.n.	69.9	11.7	6.5	6.5

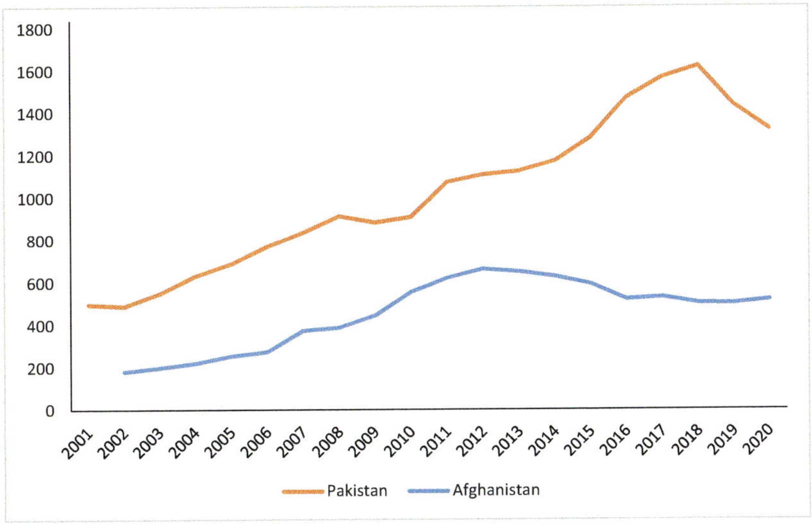

Fig. 3.5. Per capita GDP by country. (Based on World Bank figures from 2020)

Socioeconomic Context and Human Development

Income and poverty

The Human Development Index (HDI) measures achievement in human development through three dimensions: (i) health (assessed by life expectancy at birth); (ii) education (assessed by the mean years of schooling for above 24-year-old adults and the expected years of schooling for children of school-entering age); and (iii) standard of living (assessed by the gross national income per capita).

The last *Human Development Report* (UNDP, 2020) using data from 2019 ranks Afghanistan at 169 and Pakistan at 154 with HDIs of 0.511 and 0.557 (Table 3.3). Despite an improvement during the last two decades (2002–2020), Afghanistan is lagging in all socio-economic development indices compared to other countries in the region. In terms of gross national income (GNI) per capita, the south-Asian average is three times greater than that of Afghanistan and 23% higher than that of Pakistan.

In 2019, Afghanistan had a gross domestic product (GDP) of US$ 19.8 billion and the GDP per capita stood at US$ 508 (World Bank, 2022b). GDP growth for 2019 was 1.5% but –1.9% in 2020 because of the COVID-19 pandemic. The country´s gross international reserves stood at US$ 9.7 billion, equal to 16 months of imports (World Bank, 2022a).

The total GDP of Pakistan is US$ 263.7 billion with a per capita GDP of US$ 1190. The GDP growth rate for 2021 was estimated at 3.9%, which was higher than the targeted growth of 2.1% for 2020. The trend in per capita GDP fluctuation for Pakistan and Afghanistan over 20 years is shown in Fig. 3.5 (World Bank, 2022c).

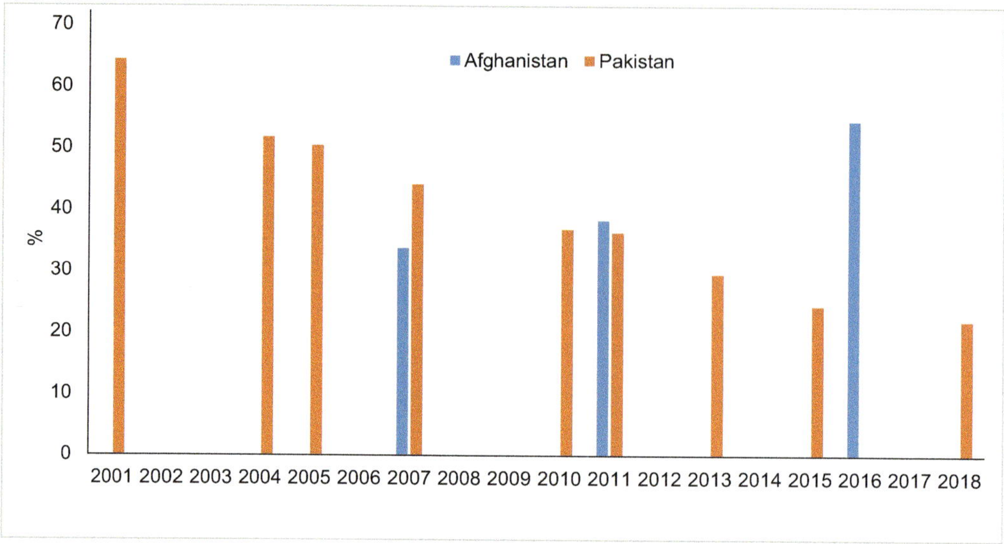

Fig. 3.6. Poverty headcount ratio at national poverty lines as a percentage of the population. (Based on World Bank figures from 2020).

In Afghanistan, since 2000, the poverty rate has never been below 33%. Poverty increased after 2010 with 54.5% of the population below the national poverty line in 2016 (Fig. 3.6). With the collapse of the Republic on 15 August 2021 and the Taliban takeover, conditions may further deteriorate. The UNDP warned in September 2021 that 97% of the population was at risk of falling below the poverty line by mid-2022 unless measures were taken by the international community to resolve the political and economic issues.[1] UNDP (2021) measured an economic contraction of 20% within a year, which may reach 30% in the following years because of the sudden halt in external financial support to social sectors. On a regional level, the highest rates of poverty were observed in the south-western region, followed by the north-east overlapping the Kabul river basin.

In Pakistan, the trend showed a continuous decrease in poverty during 2000–2020 and the estimate for 2020 was that 22% of the population would be below the national poverty line (Fig. 3.7, World Bank, 2022a). At the regional level, poverty rates are higher in the north-east and south-west of the country, while the central parts, including the Kabul, Kurram and Gomal basins, are slightly better off.

Besides poverty and human development, both Afghanistan and Pakistan have relatively poor Gender Inequality Indices (Figs 3.1 and 3.7). As per 2020 data, Pakistan ranks 135 out of 162 countries, while Afghanistan is 157 (UNDP, 2020). In Pakistan, the government seems committed to improving health, empowerment and access to the labour market for women and achieving SDG 5, and has promoted entrepreneurial opportunities for women since 2019. In Afghanistan, since 2021, the little progress made towards gender equality seems to have all but stopped.

Food Security and Nutrition

High food prices, reduced income, conflict, COVID-19 impacts, and the expected impacts of La Niña are key drivers of food insecurity in all three basins. In Afghanistan, the inflation rate in 2020 was 5.6%; 10% for food items and 1.4% for non-food items (NSIA, 2021). Since the Taliban takeover, food and fuel prices, for petrol, diesel and liquefied gas, have risen (*India*

Fig. 3.7. Girls playing, Gomal Zam area, Pakistan. (From: Khadija Begum, IWMI-Pakistan)

Table 3.4. IPC acute food insecurity phase classification between January and March 2022 in Kabul, Kurram and Gomal river basins in Afghanistan and Pakistan. (Based on IPC data for Afghanistan and Pakistan (IPC, 2021a).

		Phase 3 (Crisis)		Phase 4 (Emergency)	
Country	Basin	# People	% of population	# People	% of population
Afghanistan	Kabul basin	6,843,000		3,950,000	
Afghanistan	Kurram basin	1,055,000		384,000	
Afghanistan	Gomal basin	388,000		166,000	
Pakistan	Kabul, Kurram and Gomal basins	1,185,66	24%	285,638	6%

Calculations based on province-level data. Entire provinces are considered if only a share of the area is within the river basins.

Today, 2021). The inflation data for Pakistan shows that Consumer Price Index inflation in Pakistan increased by 9% on a year-over-year basis in 2021 and food prices went up by 10.8% for urban consumers and 9.1% for rural consumers.

In Afghanistan, 23 million people, or 55% of the population, are experiencing high levels of acute food insecurity due to conflict, COVID-19, high food prices and rampant unemployment (Integrated Food Security Phase Classification, IPC Phase 3 or above) (IPC, 2021a). Between January and March 2022, some 8.7 million people were projected to be in emergency IPC Phase 4, the highest number in the world (OCHA, 2021). In the provinces covered by Kabul, Kurram and Gomal basins, 10.8 million, 1.5 million and 554,000 people, respectively, are food-insecure and need assistance (Table 3.4).

Based on the Global Hunger Index (GHI) for 2021, Pakistan has been placed 92 out of 116 countries. Although better positioned compared to other countries such as India (ranked 101) and Afghanistan (103), the hunger situation is still at a serious level with a score of 24.7. The IPC data is not available for all districts in

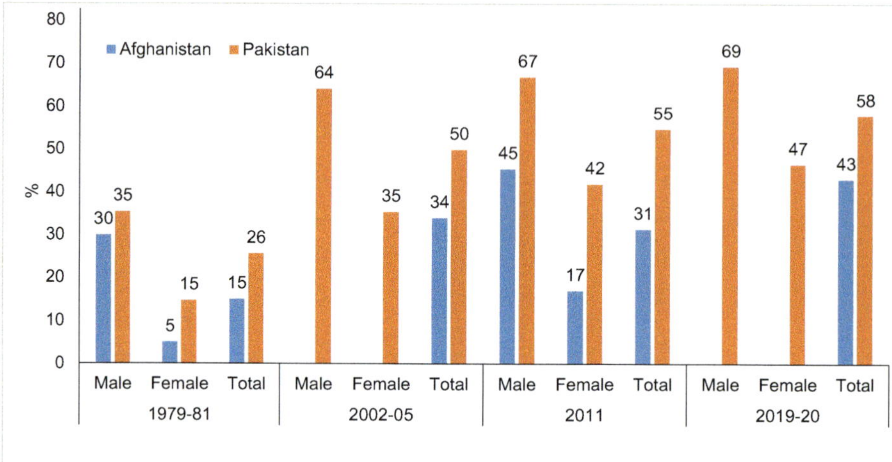

Fig. 3.8. Literacy rate among the population aged 15 and older by gender in Pakistan and Afghanistan. (Based on UNESCO, 2020, 2022).

Table 3.5. Literacy rates in Kabul, Kurram and Gomal river basins in Pakistan. (Based on data from Pakistan Bureau of Statistics census data (PBS, 2017))

		Rural			Urban			
Basin		Male (%)	Female (%)	Subtotal (%)	Male (%)	Female (%)	Subtotal (%)	Total (%)
Gomal	Illiterate	22.6	30.8	53.4	1.8	3.1	4.9	58.3
	Literate	23.8	7.8	31.5	6.0	4.0	10.2	41.7
	Total	46.4	38.5	85.0	8.0	7.1	15.0	100
Kabul	Illiterate	14.3	27.7	42.0	2.5	4.6	7.1	49.2
	Literate	25.6	12.3	37.9	8.0	5.0	13.0	50.8
	Total	39.9	40.0	79.9	10.4	9.7	20.1	100
Kurram	Illiterate	11.5	30.5	42.0	1.3	3.0	4.3	46.3
	Literate	30.6	13.6	44.2	5.7	3.8	9.5	53.7
	Total	42.6	44.1	86.24	6.96	6.80	13.76	100

the basin areas; however, using the districts in Khyber Pakhtunkhwa province (KPK) as a proxy for the whole basin region, we consider that 24% of the population in the three basins is estimated to be in IPC Phase 3 (Crisis) and 6% in IPC Phase 4 (Emergency) (IPC, 2021b).

Education

In terms of literacy, both Pakistan and Afghanistan have made substantive progress in the last 20 years. A total of 58% of the adult population is literate in Pakistan and 43% in Afghanistan as per the latest available estimates (Fig. 3.8).

District disaggregated data for Pakistan highlights that the literacy status of the population in Kabul, Kurram and Gomal basins is not uniform (Table 3.5). While the Kurram basin has the higher literacy rate, gender and rural/urban divides can be seen across the three basins.

The education sector in Afghanistan improved significantly from 2000 to 2015 and can be viewed as a success of the post-2001 period. Enrolment increased from 0.8 to 8 million in the last 20 years (World Bank, 2016). According to a 2019 survey conducted by the Ministry of Education, there were approximately 16,400 schools. The total number of students in primary, secondary, tertiary, teacher training, vocational and religious schools was 9,589,000 of which 3,686,000 were girls and women (38%) (NSIA, 2021).

School enrolment in the Kabul, Kurram and Gomal basins trends upward by including Kabul district and is proportionally higher than in the rest of the country. Fifty-five per cent of the overall number of students attending a school in Afghanistan is enrolled in these three basins (28% of the total population). In terms of literacy, the rate increased from 31% in 2002 to 43% in 2020 (NSIA, 2021) (Fig. 3.8).

For higher education, the total number of students enrolled in state-owned higher education institutions plus private higher educational institutions was 388,000, with 109,000 girls and women (28%). The number of state-owned and private universities and higher education institutions was 167 in 2020 (NSIA, 2021), of which 39 were privately run universities or higher education institutions. The number of lecturers and professors in state and private universities was 19,000, 2600 being women (14%) (NSIA, 2021).

In Pakistan, despite signs of progress toward the SDGs, almost 23 million children are out of school, including 12 million girls. This large number of children is at high risk in terms of access to educational opportunities, which is an important human right for all (GoKP, 2020).

Table 3.6 shows the net enrolment rate of boys and girls in Kabul, Kurram and Gomal basins in Pakistan in both public and private schools, including religious schools. These figures highlight the gap between girls and boys. For example, in the Kabul basin, the enrolment rate is 74% for boys and 54% for girls.

Health

Despite massive progress made by Afghanistan since the 1990s, there are still differences in health outcomes between Afghanistan and Pakistan. This is exemplified by considering life expectancy at birth and the share of cohort survival to age 65 (Fig. 3.9). The ten years of difference in life expectancy in 1990 have now been reduced, but

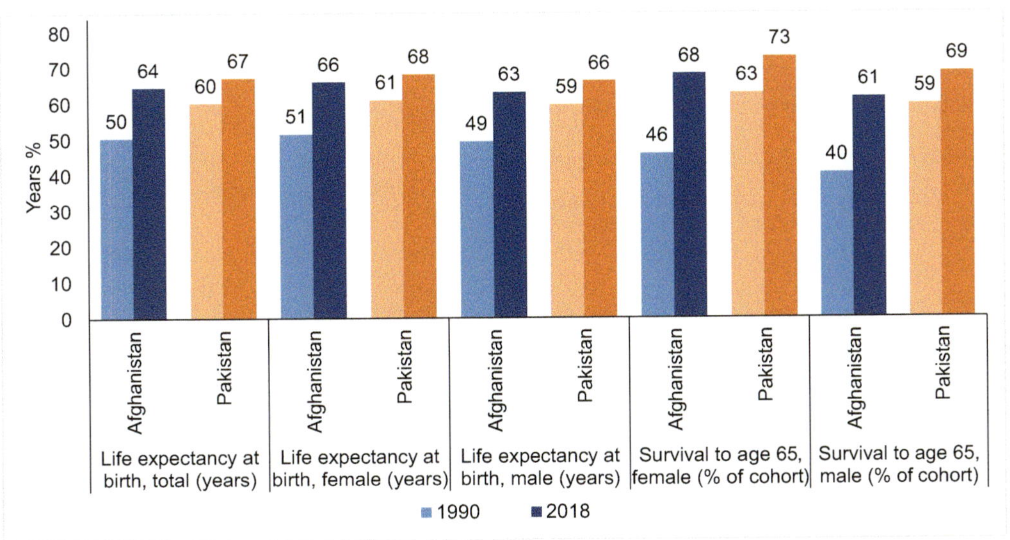

Fig. 3.9. Life expectancy to age 65 in Afghanistan and Pakistan. Numbers indicate the percentage of the total population. (Modified from World Bank Group World Development Indicators 1990, 2018 for Afghanistan and Pakistan; World Bank, 2022a)

life expectancy at birth is still 2.6 years higher in Pakistan compared to Afghanistan. Similarly, 68% of a cohort of women reaches the age of 65 in Afghanistan while almost 73% do so in Pakistan. For men, the survival rate to the age of 65 is 61% in Afghanistan and 69% in Pakistan (World Bank, 2022a).

The difference in health outcomes is a function of several factors. Secondary statistics indicate a weaker health system in Afghanistan compared to Pakistan in terms of the health service infrastructure and the number of physicians. In 2018, there were less than four hospital beds per 10,000 persons in Afghanistan and 6.3 in Pakistan (World Bank, 2022a; NSIA, 2021). The difference is even larger in terms of the number of physicians, with over three times more physicians per 10,000 person in Pakistan. Child immunization and prenatal care for women are both lower by more than ten percentage points in Afghanistan than in Pakistan.

Data at the provincial level for Afghanistan show that the share of health services is slightly higher in the provinces in the Kabul, Kurram and Gomal basins compared to the rest of the country, but this is mainly driven by the greater number of services available in Kabul district. Kabul district accounts for 37% of the medical personnel in the entire country. Sixty per cent of the public hospitals and 54% of the public hospital beds are in the three basins (Table 3.7).

Table 3.6. Net enrolment rate at schools (public and private) in Kabul, Kurram and Gomal river basins in Pakistan. (Based on data from GoP, 2020)

Basin	Girls (%)	Boys (%)	All children (%)
Gomal	37	67	52
Kabul	54	74	64
Kurram	51	86	68

For Pakistan, we used the data from KPK as a proxy for the three basins. The health services in the province comprise 122 hospitals, 371 civil dispensaries, 86 rural health centres, 782 basic health units, 37 TB clinics, 26 sub-health centres, 57 mother and child health centres, and 34 leprosy clinics (PBS, 2017). The density of these services is lower than in the rest of the country (PBS, 2021).

Water, sanitation and hygiene (WASH) conditions

Access to safe water and sanitation is a basic human right. Sustainable Development Goal 6 (SDG 6) targets water, sanitation and hygiene for all. In response to the commitment to achieve the SDGs by 2030, the government of Pakistan is keen to provide these services, particularly in rural areas. Thirty-six per cent of the population has access to safe drinking water in Pakistan and 28% in Afghanistan. For safe sanitation services in Afghanistan, 50% of the population has access and in Pakistan it is 68%. In both countries, despite improvements over the last decade, the situation is still critical and a root cause of diseases, in particular for children and women. In addition, while statistics on sanitation are encouraging, access to safe drinking water remains too low and signs of progress are too slow (Table 3.8).

National Economic Contexts

Economic growth and income

Between 2002 and 2012, Afghanistan experienced high economic growth rates of up to 21% (2009), which was mainly because of the inflow of billions of US dollars in international aid. In

Table 3.7. Health infrastructure and services. (Adapted from NSIA, 2021; World Bank, 2022a)

	Afghanistan		Pakistan
	Total	Gomal, Kabul and Kurram river basins	Total
Hospital beds (per 10,000 people)	3.9		6.3
Hospital beds in public hospitals (per 10,000 people)	1.1	1.3	
Physicians (per 10,000 people)	2.8		9.8

Table 3.8. Conditions of drinking water and sanitation in Afghanistan and Pakistan. (Adapted from UNICEF, 2021)

Type of drinking water services	Afghanistan (%)	Pakistan (%)
Proportion of population using safely managed drinking water services	28	36
Proportion of population using basic drinking water services	48	54
Proportion of population using at least basic drinking water services	75	90
Proportion of population using limited drinking water services	1	4
Proportion of population using unimproved drinking water sources	15	4
Proportion of population using surface water	9	2
Type of sanitation services		
Proportion of population using at least basic sanitation services	50	68
Proportion of population using limited sanitation services	11	11
Proportion of population using unimproved sanitation facilities	28	14
Proportion of population practising open defecation	11	17

the 2002–2010 period, foreign aid accounted for 55–90% of GDP, whereas domestic revenue contributed to a maximum of 10%. This trend has begun to wane. In 2020, the share of foreign aid accounted for 40% of the GDP (UNDP, 2021), whereas the highest share of domestic revenues in GDP reached only 14% in 2019. In recent years, following the withdrawal of international security forces and the deterioration of the security situation, coupled with the contested presidential elections in 2014, economic growth declined steadily with a minimum of –2.4% in 2020 due to the COVID-19 pandemic.

The economic growth of Afghanistan is also characterized by irregularities (Fig. 3.10), which are largely driven by climatic shocks affecting the agricultural sector (World Bank, 2022a). The years 2011 and 2018 were classified as severe drought years and coincided with falls in economic growth, which catch up the following year. The reduction in foreign aid had a buffer effect, but the worsening of climate change impacts further exacerbates the risk of these economic falls in the years to come.[2]

A major share of GDP by economic activity in 2020 was services (59%) due to the aid-induced economy, whereas the share of agriculture was 28% and that of industry was 13% of GDP (World Bank, 2022a). Despite huge deposits of minerals, Afghanistan could not generate substantial revenues from this sector during the past decades. Instead, criminal networks and insurgent organizations used mining as an income source, which, together with opium, accounted for 7–11% of GDP.

Total domestic revenues in 2020 generated from taxes, customs and non-tax items stood at USD 2.2 billion (Afghan afghani 171 billion), accounting for around 11% of GDP. The total revenues collected in the Kabul, Kurram and Gomal basins was USD 0.45 billion (AFN 34.48 billion), accounting for around 20% of the total revenues collected by the government in the year 2020 (World Bank, 2022a).

In Pakistan, high growth rates were registered in the last decade with a GDP growth rate above 5% for three consecutive years in 2016, 2017 and 2018. A slowdown was observed in 2019 and was further exacerbated by the COVID-19 pandemic in 2020. The structure of the economy may seem similar to that of Afghanistan, with 53% of the value-added being produced in services, 23% in the agricultural sector and 17% from industry in 2020. However, the figures for Afghanistan reflect aid dependency and reliance on agriculture to a greater extent than in Pakistan. In Pakistan, the share of each sector has remained constant over the last 20 years.

Beyond growth rates and the contribution of each sector to the economy, the major divergence between the GDPs in Afghanistan and Pakistan is their aggregate values: the value of the Pakistani GDP was ten times the Afghan GDP in 2020.

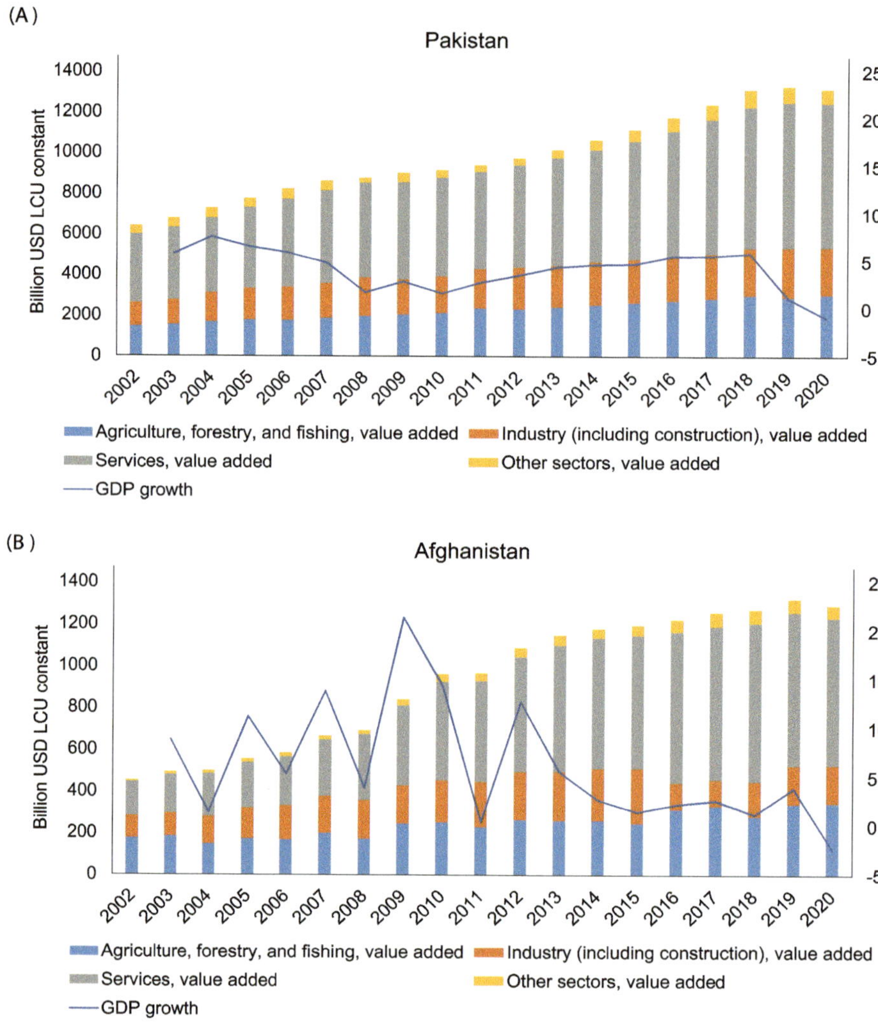

Fig. 3.10. GDP annual growth rate and value added by sector for (A) Pakistan and (B) Afghanistan. (LCU = local currency unit.) (Based on World Bank, 2022a)

Employment

In Afghanistan, between 2002 and 2012, the employment rate increased in the public and private sectors. Growth was driven by opportunities in the foreign aid service sector. However, this trend slowed during the period of transition (2012–2014) and caused a huge crisis in the job market.[3] Four out of five adult male jobs, equal to 1.9 million jobs created during the pre-transition period, were lost in 2013–2014 (World Bank, 2017). The issue was acute for youth in rural areas where half of the unemployed in 2013–2014 were below age 25 and nine out of ten jobs lost were in rural areas (Afghanistan Ministry of Economy, & World Bank, 2017). In 2018, the ILO reported that Afghanistan had the highest unemployment rate in the world estimated at between 25–30% (Tolonews, 2018)[4] for the entire adult population and that of youth unemployment (below 25 years) at 46%. Out of the 500,000 unemployed young Afghans, 72% were living in rural areas. Some estimates considered that full employment of the growing labour force in the country would require an annual growth rate of 8% (World Bank, 2017). Considering that 400,000 Afghans enter the job market annually, youth unemployment will continue to impact economic issues and fuel future conflicts instigated by terrorist networks and organizations (World Bank, 2017). Already before 15 August 2021, an increase in unemployment was forecasted, but with the takeover by the Talibans and the current economic situation, the estimated unemployment figures have increased even more.

The total number of government employees in 2020 was 414,900, of which 26% were women (NSIA, 2021). In rural areas, the proportion of women in the labour force was estimated at 29%; 60% of women work in the livestock subsector (Leao *et al.*, 2018).

In Pakistan, youth employment is a key component of the economy. Of the total population, 18% are in the 15–24 age category. This huge population of youth can be turned into a productive asset for the country if properly capitalized. The current youth unemployment rate was 11.1% in 2017–2018 (GoP, 2018). To enhance the youth employment rate, the government started a number of programmes such as the Prime Minister's Youth Entrepreneurship Scheme and the Prime Minister's Hunarmand Programme – Skills for All (GoP, 2021), the National Internship Programme, and various projects under the China Pakistan Economic Corridor Project for setting up over 100 job placement and vocational counselling centres. These programmes support skills enhancement and employment opportunities for all youth in the country including in the Kabul, Kurram and Gomal basins.

In addition to the longer-term trends, the COVID-19 pandemic affected the employment rate both in Afghanistan and Pakistan. For Pakistan, before the pandemic, approximately 56 million people were working. This number declined to approximately 35 million in the wake of COVID-19 but reverted to nearly 53 million during the 2021 fiscal year as a result of government incentives to boost employment in construction, industry and agriculture (GoP, 2021).

Key Economic Sectors in the Kabul, Kurram and Gomal River Basins

Agriculture

In Afghanistan, agriculture makes up 27% of GDP and is considered the most stable income-generating sector. The sector faces a number of challenges including water shortages, climate risks, lack of access to markets and conflict (Omerkhil *et al.*, 2020). Reviving the agricultural sector through water resources development and effective management is crucial to reducing the current aid-dependency and ensuring economic growth, food security and poverty reduction efforts (World Bank, 2010; NSIA, 2021).

For more than 80% of the population and around 90% of the poor in Afghanistan, agriculture is the main source of livelihood. About 50% of the population generates income from this sector and agriculture employs 40% of the labour force (World Bank, 2014). Compared to mining and other sectors with large development potential, agriculture has the potential to create more longer-term jobs, raise labour productivity, alleviate poverty and ensure food security in rural areas (World Bank, 2014).

In Afghanistan, 12% of the country is considered arable and only half is annually cropped while the other half remains fallow or is used as pasture land. Irrigated areas produce 85% of all agricultural production (Akhtar, 2017). It is estimated that Afghanistan has the potential to double its irrigated area from 2 million ha to 4 million ha if the existing infrastructure can be rehabilitated and investments are made in new infrastructure (World Bank, 2014). As an example, an increase of 10% of the irrigated wheat area would provide an additional US$ 1.58 billion to GDP annually (World Bank, 2014).

In the Kabul river basin, around 9% of the total drainage area is cultivated. In the downstream area of the basin, the dominant cropping pattern is wheat in rotation with maize and rice and occasionally vegetables (Favre and Kamal, 2004; World Bank, 2013). In the upstream region, the focus is on horticulture (grapes, peaches and apples). These disparities

Table 3.9. Rainfed and irrigated wheat area and production in the Kabul, Kurram and Gomal river basins in Afghanistan. (Authors' calculations based on National Statistics and Information Authority Statistical Yearbook 2020 for Afghanistan)

	Total		Rainfed		Irrigated	
Basin	Production ('000 tons)	Area ('000 ha)	Production ('000 tons)	Area ('000 ha)	Production ('000 tons)	Area ('000 ha)
Gomal	72	42.3	0.3	0.3	71.7	42
Kabul	1401	633.4	155	127.2	1245.9	506
Kurram	204	102.1	0.5	0.5	203.5	101.6

Fig. 3.11. A couple working in an agriculture field, Gomal Zam area, Pakistan. (From: Khadija Begum, IWMI Pakistan)

reflect climatic and elevation differences. The main sources of irrigation in the Kabul basin are streams, springs, karez and extraction from groundwater wells (Akhtar, 2017).[5] Table 3.9 shows that almost the entire area producing wheat is irrigated in the three basins. Yields tend to be higher in the Kurram and Kabul basins (1.99 t/ha and 2.2 t/ha, respectively) compared to the Gomal (1.7 t/ha).

In Pakistan, agricultural development is essential for economic growth and food security. Agriculture is the backbone of the economy and the main contributor to the country's GDP (18.5%) and employs 38.5% of the labour force (Fig. 3.11). Despite the economic slowdown, the performance of agriculture remained encouraging during 2020–2021 with a growth rate of 2.7% (GoP, 2021). The production of major crops showed considerable improvement during 2020 and surpassed the production targets for the current fiscal year. This occurs in a context where the government is keen to develop the agricultural sector by providing more incentives to farmers through subsidies, credits and extension programmes.

As for Afghanistan, the important crops are wheat, rice, maize, sugarcane and cotton. Wheat is an important winter crop and was sown in 2020–2021 on 8800 ha and accounts for 9.2% of the value-added in agriculture and 1.8% of GDP. The cotton crop is vital to the textile industry, which is a key sector of the economy. Cotton contributes around 0.6% to GDP and 3.1% of the value-added in agriculture. During 2020–2021, cotton was cultivated on 2080 ha with a total production of 7.1 million bales compared to 9.2 million bales in the previous cropping year. Conversion to sugarcane cultivation explains this reduction in cotton production. Rice is an important staple and cash crop. It is the second main staple food crop after wheat and the second major exportable commodity after cotton. It contributes 3.5% of value-added in agriculture and 0.7% to GDP. Nationally, rice production comprises basmati (fine) and coarse types but mostly coarse rice is produced in the transboundary river basins. In 2020–2021, rice was cultivated on 3300 ha, with a record production of 8.4 million tons against 7.4 million tons in 2019–2020. Water is one of the main limiting factors to agricultural development in Pakistan.

Industry and mining

For Pakistan, industry contributes only modestly to the economy of the territories in the three basins. Industries are located mostly in KPK with the largest share in Peshawar district. As per the latest Census of Manufacturing Industries (PBS, 2013), there were 575 industrial establishments in the three basins (Table 3.10), corresponding to 9% of the national total. The number of employees in the industrial sector is also limited and less than 5% of the population in the three basins is employed in the industrial sector. Similarly, in terms of the value of production and value-added, the industrial sector in the basins accounts for 3.7% of the national production value and 5.7% of the national value-added.

Metallurgy is the most frequent type of industrial establishment (30.1%). Agrifood industries come second, with 19.3% of the industrial establishments on the Pakistani side of the three basins. In third position, the textile industry is the main occupation of 9.3% of the firms from the three basins but is essentially located in Swabi and Swat districts (Kabul river basin).

Data on the mining sector are not published at the district level for Pakistan but only at the provincial level (GoP, 2009). Therefore, instead of considering all the districts in the basins, we consider only KPK, which covers the largest share of the three-basin area. For example, limestone and marble are also extracted in this province. While 32.6% of the mining and quarrying establishments are in KPK, they generate only 3.6% of the national value-added from all mineral industries. This suggests that the mines in KPK are less lucrative than those located in other parts of the country.

In Afghanistan, even if manufacturing played a significant role in the country's economic growth before 2021, the sector's size remains limited. Agrifood, cotton and clothing, cement, paper and printed products are the primary industries (NSIA, 2021). The industrial sector in Afghanistan is driven by the private sector and non-negligible contributions from the public sector with government-owned industries (15% of the total production value in 2021). Recent data at the provincial and district level are not available, and many factories were damaged by conflicts and are only partially operational. Malikyar (2017) provides a list of the 50 significant factories and plants in the country and we note that 23 (46%) are (or were) within the basins. Most are actually in and around Kabul but some industries were also found in Parwan and Kapisa provinces and Baghram district.

Afghanistan has vast deposits of natural resources and extraction could go beyond coal, sand, marble, gas and salt. A 2019 report from the Ministry of Mines and Petroleum (inaccessible but cited by Al Jazeera in 2021) considered that the country has over US$ 1 trillion worth of mineral resources and metals. Some of these metals and industrial minerals are in the upstream portion of the Kabul river (gems,

Table 3.10. Industries in Kabul, Kurram and Gomal river basins in Pakistan. (Authors' calculations based on Pakistan Bureau of Statistics Census of Manufacturing Industries 2005–2006 (PBS, 2013))

	Country	Number industrial establishments	Share of national number of industrial establishment	Average number of persons engaged daily	Share of national number of persons engaged	Value of production (in '000 PKR)	Share of national value of production	Census value added at producer prices (in '000 PKR)	Share of national value added
Gomal basin	Pakistan	23	0.004	2109	0.002	2,271,995	0.001	480,456	0.000
Kabul basin	Pakistan	511	0.080	40,477	0.043	95,998,741	0.033	54,336,453	0.049
Kurram basin	Pakistan	41	0.006	2764	0.003	10,618,445	0.004	7,723,486	0.007
Total three basins	Pakistan	575	0.090	45,350	0.048	108,889,181	0.037	62,540,395	0.057
Total country		6417	1.000	941,283	1.000	2,929,319,774	1.000	1,103,414,202	1.000

*1USD = 226 PKR at time of writing (July 2022).

Table 3.11. Electricity production and imported electricity in Afghanistan (2018–2020).

	Electricity production (GWh)		
Type of energy	2018	2019	2020
Hydro	786	1088	881
Thermal	151	128	81
Diesel	42	49	38
Solar	-	21	32
Wind	-	-	-
Nuclear	-	-	-
Bagasse/biogas	-	-	-
Public sector GENCOs	-	-	-
SPPs/CPPs	-	-	-
Domestic production	979	1286	1031
Renewable production	786	1109	913
Fossil-fuel production	193	177	118
Country	Electricity imports (GWh)		
	2018	2019	2020
Iran	743	758	817
Tajikistan	914	1462	927
Turkmenistan	737	688	733
Uzbekistan	2592	2023	2675
Imported electricity	4986	4932	5152
Available electricity	5965	6217	6183
Imported electricity (%)	84	79	83
Domestic production (%)	16	21	17
Renewable production (%)	80	86	89
Fossil-fuel production (%)	20	14	11
Solar production (%)	-	1.6	3.1
Hydro production (%)	80.3	84.6	85.4

lithium and caesium in Nuristan district; magnesite and talc in Nangarhar; copper, chromite and cobalt in Kabul, Logar and Wardak provinces; emeralds and silver in Panjshir province).

Power sector

Afghanistan has vast resources for electricity generation, in particular from renewable energy sources. According to the Ministry of Energy and Water (2015), Afghanistan's hydro potential is estimated at 23 GW and wind at 66 GW. The solar potential is estimated to be 222 GW. The country also has the potential to generate power from biomass and geothermal resources. In addition, Afghanistan has considerable natural gas resources in three basins, the Amu Darya, Helmand and Katawaz (Asian Development Bank, 2016).

Despite this power-generation potential, the per capita consumption rate of electricity is 195 kWh/a (Ministry of Energy and Water, 2015), which is one of the lowest in the world; 83% of the available electricity is imported from Central Asia and Iran (Tables 3.11 and 3.12). From the 1960s to 1988, Afghanistan was exporting its natural gas to the Soviet Union, which contributed up to 38% to the government's annual budget (Bezhan, 2018). However, during the period 1978–2001, the power infrastructure was badly damaged (World Bank, 2010). Since 2002, Afghanistan has rehabilitated and expanded its electricity network and increased the connection rate from less than 5% to 35%.

Hydropower is the main source of production and recent public–private partnerships have been focusing on solar and gas-to-power to increase production capacity. However, Afghanistan is lacking a unified grid (Afghan Power Sector Master, 2013) resulting in huge inefficiencies and network losses. Recent reports indicate power outages of up to 12 hours in major cities are having an impact on private households and commercial activities.

As expected, the Kabul river basin plays a central role in electricity production. Sixty-six per cent of the country's available hydropower capacity (222 MW) is there (Table 3.13). To address the country's dependency on imported electricity, the increased demand for power and

Table 3.12. Electricity production in Pakistan (2018–2020). (Authors' calculations based on NEPRA, 2020; Unwin, 2020; NSIA, 2021)

	Electricity production (GWh)	
Type of energy	2018–2019	2019–2020
Hydro	33,096	38,988
Thermal	62,598	60,720
Diesel	-	-
Solar	715	705
Wind	3232	2882
Nuclear	9006	9705
Bagasse/biogas	894	564
Public sector GENCOs	13,017	7908
SPPs/CPPs	405	171
Domestic production	122,963	121,643
Renewable production	46,943	52,844
Fossil-fuel production	76,020	68,799
Country	Electricity imports (GWh)	
	2018–2019	2019–2020
Iran	487	514
Tajikistan	-	-
Turkmenistan	-	-
Uzbekistan	-	-
Imported electricity	487	514
Available electricity	136,532	134,746
Imported electricity (%)	0.4	0.4
Domestic production (%)	99.6	99.6
Renewable production (%)	38.2	43.4
Fossil-fuel production (%)	61.8	56.6
Solar production (%)	0.6	0.6
Hydro production (%)	26.9	32.1

Table 3.13. Installed hydroelectricity dams in Kabul, Kurram and Gomal river basins. (From: for Pakistan, WAPDA, 2012; NEPRA, 2020. For Afghanistan: Afghan Power Sector Master, 2013; Afghanistan Inter-ministerial Commission for Energy, 2017)

	Hydropower plants	Installed capacity (MW)
Afghanistan	Naghlu	100
	Mahipar	66
	Sarobi	22
	Darunta	11.5
	Pul-e-Khomri I & II	12.91
	Chak-e-Wardak	3.3
	Jabul Seraj	2.5
	Charikar	2.4
	Assassab/Kunar	0.7
	Ghorband	0.3
	Total installed hydroelectricity capacity in Kabul basin	222
	Total installed hydroelectricity capacity in Afghanistan	336
	Share of installed hydroelectricity capacity in Kabul basin (%)	66
Pakistan	Gomal basin (Gomal Zam Dam)	17.4
	Kabul basin (Warsak Dam)	243+282*
	Kurram basin (Kurrum Tangi Dam)	83.4
	Total installed hydroelectricity capacity in the 3 basins (selected dams only)	343.8
	Total installed generation capacity in Pakistan	38,719
	Share of hydroelectricity installed capacity in 3 basins (%)	0.8

*Planned upgrade.

water for irrigation and household purposes, the 2002–2021 Afghan government began to explore new projects in the Kabul river basin (World Bank, 2010; Afghan Power Sector Master, 2013).[6]

The power sector in Pakistan is drastically different from that of Afghanistan. During the 2019–2020 financial year, total electricity generation in the country was 135,000 GWh. To respond to the demand and meet growing energy needs, 514 GWh were imported from Iran (0.4% of available electricity). According to the World Bank, about 74% of the population had access to electricity in 2019 (World Bank, 2022b).

Pakistan still relies largely on thermal energy and natural gas for electricity production. However, this dependency has been decreasing and sources are diversified with non-negligible production from hydro, nuclear, wind, solar and biogas sources. The main producers of electricity in Pakistan are the Water and Power Development Authority (WAPDA), private generation companies (GENCOs) and small-capacity power plants (SPPs/CPPs). Compared to Afghanistan, hydropower contributes a smaller share of the total national production (32%).

Conclusions

This chapter has noted the divide between and within the two countries in terms of population size, migration dynamics, economic indicators, and trends in production of, and access to, power. However, there are also clear similarities. The economic structure is similar on both sides of the border and agriculture still plays a major role in the economic life and livelihoods of the local populations. The basins are mostly rural areas where poverty prevails and gender disparities in development indicators are striking. Despite differences in nationality, the populations of the Kabul, Kurram and Gomal basins depend on their water resources and share challenges related to food security, poverty alleviation, and mitigation and adaptation to climate change.

Before concluding, it is worth acknowledging one caveat related to methods. Often, the administrative units for which the data were collected do not entirely match the geographical and hydrological units of the basins. To reconcile this, we considered here districts with the largest area within the watershed area as part of the river basin but acknowledge that there may be minor mismatches between the hydrological definition of the basin and the one followed here.

Ultimately, this chapter can be seen as a first step towards building a shared perspective on demography and human and economic conditions in the major transboundary Afghan–Pakistani watercourses. As the ultimate ends to which the means of water management contribute, it is important to characterize development conditions in the three basins. Realities outlined in the chapter suggest there remain major challenges to human and economic development, which underlines the need to use multiple levers, including basin management, to improve outcomes. Nonetheless, the conditions described in this chapter are the outcome of numerous factors which would benefit from renewed policy development and sustained investment within and across a range of sectors (e.g. agriculture, industry and energy). As water cuts across all concerned sectors, it would appear critical to consider how management of the resource can be optimized to support such efforts. Such optimization is best undertaken at a basin level even if a basin crosses borders. It is imperative to achieve some level of co-ordination so that water can best be channelled to support the desperately needed human and economic development in the Kabul, Kurram and Gomal basins.

Notes

[1] https://www.undp.org/press-releases/97-percent-afghans-could-plunge-poverty-mid-2022-says-undp (accessed July 2022)

[2] For example, the drought of 2022 has been declared the worst in the last three decades. See: https://www.humanitarianresponse.info/sites/www.humanitarianresponse.info/files/documents/files/afghanistan-humanitarian-response-plan-2022.pdf (accessed 14 September 2022).
[3] Transition-transfer of responsibility for security from international forces to Afghan forces with the withdrawal of the bulk of international forces in 2014.
[4] https://tolonews.com/business/afghanistan-has-highest-unemployed-work-force-ilo
[5] Karez is an indigenous method of irrigation in which groundwater is tapped remotely and brought to a command area through a tunnel.
[6] Baghdara, Surobi 2, Kunar A (Shal), Gulbahar, Capar and Kama were new planned projects. In total, they would add 1460 MW of capacity (Afghan Power Sector Master, 2013).

References

Afghanistan Inter-ministerial Commission for Energy (2017) Electricity import cost. Available at: https://sites.google.com/site/iceafghanistan/electricity-import-cost (accessed 15 September 2022).

Afghanistan Ministry of Economy, & World Bank (2017) Afghanistan poverty status update: progress at risk. World Bank. Available at: https://documents1.worldbank.org/curated/en/667181493794491292/pdf/114741-WP-v1-P159553-PUBLIC.pdf

Afghan Power Sector Master (2013). Available at: https://www.adb.org/sites/default/files/project-document/76570/43497-012-afg- tacr.pdf (accessed July 2022).

Akhtar, F. (2017) Water availability and demand analysis in the Kabul River Basin. Doctoral dissertation, University of Bonn, Germany.

Asian Development Bank (2016) *TA-8401 AFG: Gas Development Master Plan*. Asian Development Bank, Manila.

Bezhan, N. (2018) *Aid Paradoxes in Afghanistan*. Routledge, London. DOI: 10.4324/9781315170701.

Cibea, A., Dearing, S., Gures, G., Rehman Khan, A. and Wolfsgruber, I. (2013) *Pakistan Migration Country Report*. International Center for Migration Policy Development, Vienna.

Favre, R. and Kamal, G.M. (2004) *Watershed Atlas of Afghanistan: Kabul*. Government of Afghanistan, Ministry of Irrigation, Water Resources and Environment, Kabul, Afghanistan.

GoKP (Government of Khyber Pakhtunkhwa province) (2020) *Annual School Census Report for Settled Districts 2020–2021*. Elementary and Secondary Education Department, Government of Khyber Pakhtunkhwa. Available at: http://175.107.63.45/NewIMUSite/images/reports/ASC_Report_2020-21_Final.pdf (accessed July 2022).

GoP (Government of Pakistan) (2009) Census of mining and quarrying industries (CMQI) 2005–2006. Government of Pakistan, Statistics Division, Federal Bureau of Statistics, Islamabad. Available at: https://www.pbs.gov.pk/sites/default/files/industry_mining_and_energy/publications/cmq200506/Mining%20Report%20Final.pdf (accessed January 2022).

GoP (2018) *Labour Force Survey 2017–2018*. Pakistan Bureau of Statistics, Government of Pakistan, Islamabad. Available at: https://www.pbs.gov.pk/sites/default/files//Labour%20Force/publications/lfs2017_18/Annual%20Report%20of%20LFS%202017-18.pdf (accessed July 2022).

GoP (2020) *Labour Migration Report 2020*. Government of Pakistan, Bureau of Emigration & Overseas Employment, Islamabad.

GoP (2021) *Pakistan Economic Survey 2020–2021*. Economic Adviser's Wing, Finance Division, Government of Pakistan, Islamabad. Available at: https://www.finance.gov.pk/survey_2021.html (accessed 19 July 2022).

Hussain, A. and Jan, I. (2020) Effect of foreign remittances on social festivities in 2021. Available at: https://www.thefreelibrary.com/Effect+of+Foreign+Remittances+on+Social+Festivities+in+District...-a0645999236 (accessed 15 September 2022).

IFO Institute (2021) Afghanistan's free fall – return of the Taliban and flight as a last resort. Available at: https://www.cesifo.org/de/publikationen/2021/aufsatz-zeitschrift/afghanistans-free-fall-return-taliban-and-flight-last-resort (accessed 5 January 2022).

India Today (2021) Kabul residents struggle to make ends meet as food, fuel prices soar under Taliban govt. 27 September. Available at: https://www.indiatoday.in/world/story/afghanistan-food-fuel-prices-soar-taliban-kabul-residents-struggle-1857555-2021-09-27 (accessed 25 September 2021).

IOM (International Organization for Migration) (2018) Documentation of undocumented Afghans at Afghan Citizen Card Centers. Available at: https://www.iom.int/infographics/documentation-undocumented-afghans-afghan-citizen-card-acc-centers (accessed 19 July 2022).

IPC (2021a) Afghanistan: acute food insecurity situation, March–May 2021 and projection for June–November 2021. Available at: https://www.ipcinfo.org/ipc-country-analysis/details-map/en/c/1154300/ (accessed 4 January 2022).

IPC (2021b) Integrated food security phase classification Pakistan – Khyber Pakhtunkhwa. October 2021–June 2022. Available at: https://reliefweb.int/report/pakistan/khyber-pakhtunkhwa-pakistan-ipc-acute-food-insecurity-analysis-october-2021-june (accessed 19 July 2022).

Ishfaq, S., Ahmed, V., Hassan, D. and Javed, A. (2017) Internal migration and labour mobility in Pakistan. In: Rajan, S.I. (ed.) *South Asia Migration Report 2017*. Routledge, South Asia edition.

Leao, I., Ahmed, M. and Kar, A. (2018) *Jobs from Agriculture in Afghanistan*. World Bank Group, Washington, DC. DOI: 10.1596/978-1-4648-1265-1.

Malikyar, G.M. (2017) The state of environment of Afghanistan (2010–2017): pressures, progress, challenges/gaps. Available at: https://www.researchgate.net/publication/326735751_The_state_of_Environment_of_Afghanistan_2011-2017 (accessed 3 January 2022).

Ministry of Energy and Water (2015) *Renewable Energy Policy*. Available at: http://www.red-mew.gov.af/wp-content/uploads/2014/11/Afghanistan-Renewable-Energy-Policy-English-and-Dari.pdf (accessed July 2022).

NEPRA (2020) *State of Industry Report – 2020*. National Electric Power Regulatory Authority, Islamabad, Pakistan.

NSIA (2021) *Afghanistan Statistical Yearbook.* Issue 42. Available at: https://invest.gov.af/theme3/wp-content/uploads/2021/06/Afghanistan-Statistical-Yearbook-first-Version.pdf (accessed January 2022).

OCHA (2021) Afghanistan – complex emergency fact sheet 7, fiscal year 2021. US Agency for International Development. Available at: https://cutt.ly/TEKCqHI (accessed 2 January 2022).

Omerkhil, N., Chand, T., Valente, D., Alatalo, J.M. and Pandey, R. (2020) Climate change vulnerability and adaptation strategies for smallholder farmers in Yangi Qala District, Takhar, Afghanistan. *Ecological Indicators* 110, 105863. DOI: 10.1016/j.ecolind.2019.105863.

PBS (Pakistan Bureau of Statistics) (2013) *Census of Manufacturing Industries-2005 District-Wise Report*. Available at: https://www.pbs.gov.pk/sites/default/files/industry_mining_and_energy/publications/CMI_2005-06_district-wise.pdf (accessed 3 January 2022).

PBS (2017) *Provincial Census Report for Khyber Pakhtunkhwa*. Government of Pakistan Ministry of Planning, Development and Special Initiatives. Pakistan Bureau of Statistics, Islamabad. Available at: https://www.pbs.gov.pk/sites/default/files//population_census/ncrpcr/PCR%20Khyber%20Pakhtunkhwa.pdf (accessed July 2022).

PBS (2021) *Key Findings Report, Pakistan Social and Living Standards Measurement (PSLM)*. District level survey 2019–20. Government of Pakistan Ministry of Planning, Development and Special Initiatives, Pakistan Bureau of Statistics, Islamabad. Available at: https://www.pbs.gov.pk/sites/default/files//pslm/publications/pslm_district_2019-20/PSLM_2019_20_District_Level.pdf (accessed October 2022).

UNDP (United Nations Development Programme) (2020) *Human Development Report: The next Frontier: Human Development and the Anthropocen*. Briefing note for countries on the 2020. Available at: https://hdr.undp.org/sites/default/files/Country-Profiles/PLW.pdf (accessed July 2022).

UNDP (2021) *Afghanistan Socio-Economic Outlook 2021-2022: Averting a Basic Needs Crisis*. Available at: https://www.undp.org/publications/afghanistan-socio-economic-outlook-2021-2022-averting-basic-needs-crisis#:~:text=December%201%2C%202021&text=The%20report%20indicates%20that%20the,corrective%20action%20is%20not%20taken (accessed July 2022).

UNESCO (2020) Afghanistan – UNESCO highlights key challenges for education, science and culture in the country. Available at: https://en.unesco.org/news/afghanistan-unesco-highlights-key-challenges-education-science-and-culture-country (accessed 13 January 2022).

UNESCO (2022) Education and literacy. Institute for Statistics, Afghanistan. Available at: http://uis.unesco.org/country/AF (accessed 13 January 2022).

UNHCR (2022) Afghanistan: WASH (Water, Sanitation and Hygiene): safe water, sanitation, and good hygiene practices. Available at: https://www.unicef.org/afghanistan/wash-water-sanitation-and-hygiene (accessed 3 January 2022).

UNICEF (2021) Sanitation and drinking water. Available at: https://data.unicef.org/country/afg/ (accessed 4 January 2022).
Unwin, J. (2020) What does Pakistan's energy mix look like and what is its future? Available at: https://www.power-technology.com/features/pakistan-energy-mix/ (accessed 13 January 2022).
WAPDA (2012) Kurram Tangi Dam multi-purpose project. Available at: https://web.archive.org/web/20120610151652/http://www.wapda.gov.pk/htmls/KurmTngDam.htm (accessed 19 July 2022).
World Bank (2010) Scoping strategic options for development of the Kabul river basin: a multisectoral decision support system approach. Sustainable Development Department, South Asian Region, World Bank. Available at: https://openknowledge.worldbank.org/handle/10986/18422 (accessed 3 January 2022).
World Bank (2013) Afghanistan: Emergency Irrigation Rehabilitation Project. Available at: http://www-wds.worldbank.org/external/default/WDSContentServer/WDSP/IB/2013/07/01/000445729_20130701124006/Rendered/PDF/ICR21980P078930C0disclosed060270130.pdf (accessed 5 January 2022).
World Bank (2014) Islamic Republic of Afghanistan agricultural sector review: revitalizing agriculture for economic growth, job creation and food security. Available at: https://openknowledge.worldbank.org/handle/10986/21733 (accessed 19 July 2022).
World Bank (2016) Agriculture and rural development. Available at: http://data.worldbank.org/topic/agriculture-and-rural-development?locations=AF-AD (accessed 7 January 2022).
World Bank (2017) Afghanistan poverty progress update: progress at risk. Available at: https://openknowledge.worldbank.org/handle/10986/26668?show=full (accessed 6 January 2022).
World Bank (2022a) World development indicators databank 1990, 2018. Available at: https://databank.worldbank.org/source/world-development-indicators (accessed 19 July 2022).
World Bank (2022b) Electric power consumption (KWh) – South Asia. Available at: https://data.worldbank.org/indicator/EG.USE.ELEC.KH.PC?locations=8S (accessed 10 January 2022).
World Bank (2022c) Pakistan GDP. Available at: https://tradingeconomics.com/pakistan/gdp (accessed 19 July 2022).

4 Climate

Muhammad Zia ur Rahman Hashmi, Muhammad Tousif Bhatti and Mujib Ahmad Azizi

Chapter Overview

Climate change is a major threat in all the shared river basins in South Asia owing to the huge populations they sustain and the complex regional dynamics. This chapter summarizes the available knowledge related to changing climate trends in three Afghan–Pakistani transboundary river basins: the Kabul, Kurram and Gomal. A key finding indicates a consistent rise in temperature over the last 30 years. By the end of the 21st century, temperatures may rise by an additional 3–4°C under RCP[1] 4.5 and 5–6°C under RCP 8.5, relative to 2020 levels. The potential impact of temperature rise is compounded by considerable uncertainty associated with the current and future behaviour of precipitation in the three basins. The findings in this chapter will help practitioners and policymakers visualize the nature and scope of likely climate challenges in the three basins.

Fig. 4.1. Agriculture land, Panjshir valley Kabul river, Afghanistan. (Photo: Alex Treadway, ICIMOD)

DOI: 10.1079/9781800622371.0004

Introduction

Precipitation is regarded as one of the most significant variables in climate systems because of its fundamental importance in energy cycles (Guan *et al.*, 2015). Shifts in precipitation patterns can lead to short- and long-term changes in biological, hydrologic and biogeochemical phenomena (Salehi *et al.*, 2020). Similarly, temperature changes have a significant impact on climate variability. Thus, quantifying precipitation and temperature changes is critical to understanding the climate impact on various components of an environment. While the adequate prediction of climate trends in the Kabul, Kurram and Gomal basins is a challenge due to the lack of data, broad trends can nonetheless be identified.

Collection of climate data in these shared basins began in the 20th century but periodic conflicts have disrupted data-collection activities and resulted in scattered data of variable quality. Taraky (2021), for example, used data from as far back as the mid-1930s from a network of more than 70 weather stations. In the absence of sufficient observed climate data, researchers introduced alternative ways to generate continuous climate records (Bokhari *et al.*, 2018).

This chapter reviews and synthesizes investigations and knowledge products related to changing climate trends in the three transboundary river basins shared between Afghanistan and Pakistan (Kabul, Kurram and Gomal). The chapter begins by describing the state of the climate in the Kabul river basin, including its annual and seasonal variations, and changes in evapotranspiration and the cryosphere. The authors then discuss the climate situation in the Kurram and Gomal river basins and offer climate projections and suggestions for climate governance. Analysing the available evidence on climate change and variability will help practitioners and policymakers negotiate a common understanding of climate trends, and challenges and opportunities in their shared watercourses. While there is no new analysis in this chapter, the chapter synthesizes the existing climate knowledge across the two countries to establish a basis for further knowledge production.

Kabul River Basin

Overview. Based on the dominant hydroclimatology, the Kabul river basin (Fig. 4.1) has been classified into glaciated and non-glaciated sub-basins (Gurung *et al.*, 2020). The three glaciated sub-basins are the Kunar (Chitral), Alingar and Upper Panjshir (Fig. 4.2). The five non-glaciated sub-basins include Upper Kabul, Chak Wa Logar Rod, Ghurband, Lower Kabul and Lower Panjshir. Elevation varies from more than 7000 metres above sea level (masl) to less than 1000 masl (Fig. 4.3).

Extent of monitoring in the Kabul basin. There are more monitoring stations in Afghanistan than in Pakistan (Fig. 4.3). Unfortunately, data generated through such monitoring often remain limited in terms of temporal and spatial coverage (e.g. Gurung *et al.*, 2020). There are substantial data gaps, particularly for stations in Afghanistan.

Data quality. There are 56 climate-monitoring stations in the Kabul river basin (Table 4.1). However, data from each station are generally available for less than a decade. The data parameters typically available include temperature and precipitation, which are recorded on an hourly basis.

Long-term data. Long-term climate data are a primary requirement for any robust analysis of climate change and its impacts. Where ground data (Fig. 4.4) are not readily available, remote sensing data can be used as proxy data. Options include the High Asia Refined Analysis (HAR) (Wang *et al.*, 2021), and the Climate Hazards Group Infra-Red Precipitation with Station Data (CHIRPS) used by Baig *et al.* (2020) for the Kabul river basin.

Seasonal and Annual Climate Variability and Trends

The climate of the Kabul river basin is characterized as semi-arid and continental (hot in summer and cold in winter). During the winter, the average

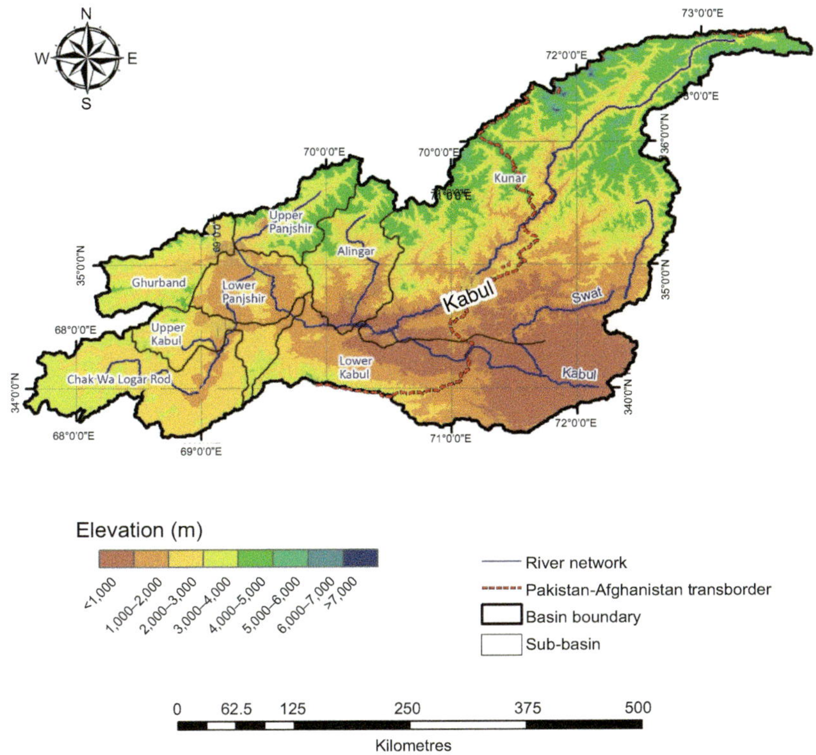

Fig. 4.2. The Kabul river basin, glaciated and non-glaciated sub-basins.

maximum precipitation is 110 mm with yearly precipitation from 350 to 800 mm (the long-term average is around 722 mm). July is the hottest month with the highest maximum temperature of around 28°C, while the average minimum temperature in winter is around 6°C. The mean annual temperature ranges from −3 to 21°C (Hassanyar and Tsutsumi, 2017; Khatiwada *et al.*, 2021). Precipitation in winter falls in relation to rain and snowfall caused by synoptic-scale eastward propagating western disturbances advected by the upper-level westerly jetstream (Yadav *et al.*, 2010).

Monthly climatology, temperature and precipitation closely related in the Kabul basin. Long-term monthly climatology at two selected stations in the Kabul river basin indicates peak temperatures in June/July and peak precipitation in March (Fig. 4.5.). Chitral represents the high-elevation mountainous part of the basin while Peshawar represents the lowest elevation with flatter terrain. The Kabul river basin has different climates in its low- and high-elevation zones as represented by Peshawar climate stations (warmer temperatures, receives precipitation both in summer and winter) and Chitral (cooler than low-elevation zones, receives most of its precipitation in winter).

Spatial diversity in climate. Multiple precipitation climatologies prevail in the topographically diverse Kabul river basin. Based on an analysis of the available observed data in Afghanistan and Pakistan, Casale *et al.* (2020) divided the basin into five precipitation zones (Shukhi, Gharu, East, West and Centre). Of these five zones, only in the west zone is there an inverse relationship seen between precipitation and rise in altitude (i.e. precipitation decreases with a rise in altitude). In all the other zones, a clear increase is seen in precipitation as we go

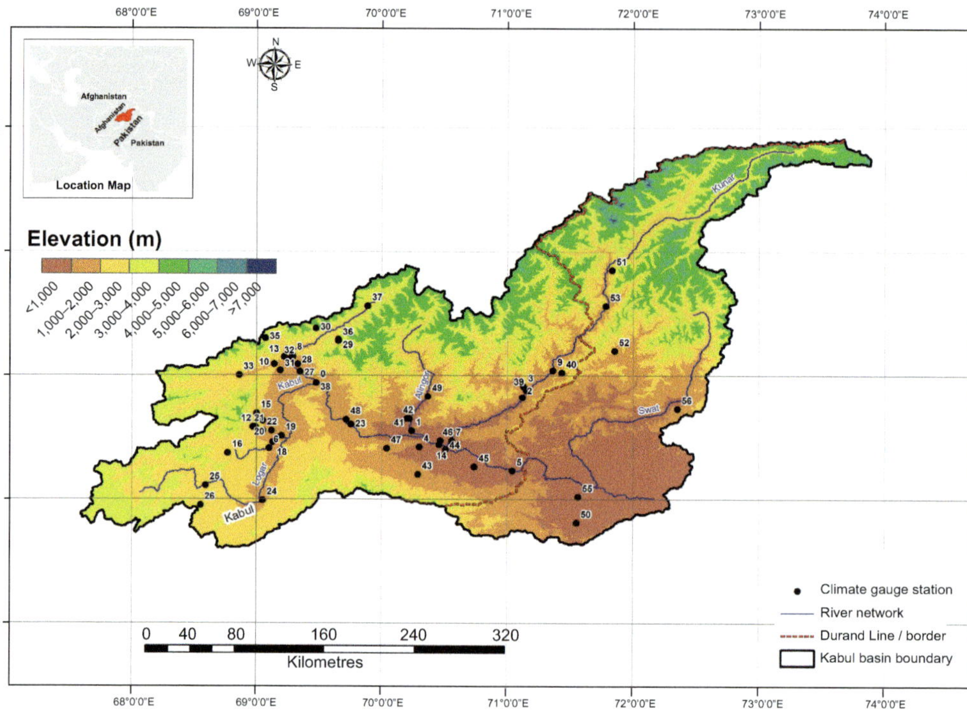

Fig. 4.3. Climate stations in the Kabul river basin.

up in terms of altitude. Fig. 4.6. (Bokhari *et al.*, 2018) shows spatial maps of the historical climatology of precipitation and temperature in the basin for the period 1975–2005 as derived by spatial interpolation (thin-plate smoothing splines method) of the *in situ* observations of the Pakistan Meteorological Department. These maps are a good depiction of the long-term average spatial pattern of temperature and precipitation across the basin in relation to elevation gradients. The figures show that going from west to east (north is upwards in all these figures), both temperature and precipitation increase. The opposite is true when going from south to north.

An Evolving Historical Climate

Detection of changes in the main climatic parameters (i.e. temperature and precipitation) in the Kabul river basin have gained some popularity in the last few years. Data quality and availability issues were the main reasons researchers have been hesitant to undertake climate research in this highly climate-sensitive region. Some studies have indicated a 2% increase in the mean-maximum temperature data for the period 1961–1990 (Sidiqi *et al.*, 2018) and a decrease in winter precipitation (Shroder and Ahmadzai, 2016; Bokhari *et al.*, 2018; Khatiwada *et al.*, 2021). There appears to be more precipitation on the Pakistan side of the basin when the climatological means of the two periods 1963–1988 and 1989–2014 are compared to detect changes in mean-maximum and -minimum temperatures and precipitation. The latter period showed an increase in all three variables, especially in winter (Yaseen *et al.*, 2020).

Floods. Recurrent flash flooding due to extreme rainfall events is quite common in these basins, especially on the Pakistan side (Hashmi *et al.*, 2020). Trend analysis of the two climate-extreme indices (i.e. maximum one-day rainfall and maximum five-day rainfall) is considered particularly useful when an assessment of flood risk is required. Through a similar approach,

Table 4.1. List of climate stations in the Kabul river basin.

ID	Station	Parameter	Length of record	Latitude	Longitude	Elevation (masl)
	Afghanistan					
0	Pul-i-Shokhi	T, P	2012–2016	34.94	69.48	1374
1	Pul-i-Qarghayi	P	2008–2016	34.55	70.24	643
2	Nawabad	P	2008–2015	34.82	71.12	796
3	Changhasarai	P	2008–2016	34.91	71.13	847
4	Sultanpor	P	2008–2016, 2012–2015	34.42	70.3	686
5	Dakah	P	2009–2016	34.23	71.04	419
6	Tang-i-Sayedan	P	2008–2015	34.41	69.1	1870
7	Pul-i-Kama	P	2008–2015	34.47	70.56	558
8	Tang-i-Gulbahar	P	2008–2015, 2013–2015	35.16	69.29	1625
9	Asmar	P	2008–2016	35.04	71.36	950
10	Pul-i-Ashawa	P	2008–2016	35.09	69.14	1624
11	Bagh-i-Omomi	P	2008–2016	35.15	69.29	1587
12	Qala-i-Malek	P	2008–2018	34.58	68.97	2211
13	Bagh-i-Lala	P	2008–2016	35.15	69.22	1698
14	Pul-i-Behesod	P	2009–2016	34.44	70.46	555
15	Shakardara	P	2009–2018	34.69	69	2168
16	Pul-i-Surkh	P	2009–2018, 2009–2017	34.37	68.77	2216
17	Badam Bagh	P	2009–2014	34.55	69.12	1803
18	Darulaman	P	2009–2014	34.46	69.13	1822
19	Gul Khana	P	2009–2014	34.51	69.2	1793
21	Paghman	P	2009–2014	34.58	68.99	2142
22	Qargha	P	2009–2014	34.55	69.05	1934
23	Sarobi	P	2009–2014	34.6	69.76	959
25	Check	P	2009–2014	34.11	68.59	2283
26	Jaghatoo	P	2009–2014	33.95	68.55	
28	Kohestan	P	2009–2014	35.09	69.33	1536
29	Dara Panjsheer	P	2009–2014	35.29	69.66	2330
30	Dashtak	P	2009–2014	35.38	69.48	3401
31	Charikar	P	2009–2014	35.04	69.19	1559
32	Jabulsaraj	P	2009–2014	35.13	69.25	1656
33	Seya Gerd	P	2009–2014	35	68.86	1848
34	North Salang	P	2009–2014	35.32	69.02	3586
35	South Salang	P	2009–2014	35.3	69.07	2982
36	Keraman	P	2009–2014	35.28	69.66	2232
37	Khawak	P	2009–2014	35.56	69.89	2405
38	Shukhi	P	2009–2014	34.94	69.48	1374

Continued

Table 4.1. Continued

ID	Station	Parameter	Length of record	Latitude	Longitude	Elevation (masl)
39	Asad Abad	P	2009–2014	34.88	71.15	875
40	Asmar	P	2009–2014	35.02	71.43	1752
41	Laghman	P	2009–2014	34.65	70.22	730
42	Mehtarlam	P	2009–2014	34.65	70.2	751
43	Agam	P	2009–2014	34.2	70.29	1205
44	Farm Jadeed	P	2009–2014	34.4	70.51	554
45	Ghazi Abad	P	2009–2014	34.26	70.74	552
46	Jalalabad	P	2009–2014	34.47	70.47	597
47	Sheshambagh	P	2009–2014	34.41	70.04	579
48	Naghlu	T, P	2009–2014	34.64	69.72	998
49	Pul-i-Nalyar	T, P	2009–2014	34.83	70.37	902
50	Cherat	T, P	1963–2014	33.81	71.55	1372
	Pakistan					
51	Chitral	T, P	-	35.85	71.83	1497.8
52	Dir	T, P	-	35.2	71.85	1375
53	Drosh	T, P	-	35.56	71.78	1463.9
55	Peshawar	T, P	-	34.02	71.56	320
56	Saidu Sharif	T, P	-	34.73	72.35	961

T, temperature;P, precipitation;

future flood risk analysis for the Pakistan side of the Kabul river basin reveals a significant increasing trend in extreme precipitation, especially the maximum one-day and maximum five-day rainfall leading to an increased risk of flooding in the entire basin, as well as urban flooding in the main cities like the capital city of Khyber Pakhtunkhwa, Peshawar. Also, in the northern part of the basin, there is a decreasing frequency of annual and monsoonal rainfall (Mehmood *et al.*, 2021).

Actual evapotranspiration variability. Inter- and intra-annual changes in actual evapotranspiration (ETa) are strongly linked with the amount of moisture available and temperature fluctuations. In regions with a semi-arid climate and little vegetation, significant fluctuations in ETa are common. Studies that provide a comprehensive analysis of changes in ETa in the three basins are hard to find. However, the available literature shows that the annual average ETa for the early part of the 21st century and the maximum ETa levels across the Kabul river basin are normally found from May to August, with magnitudes in the range of 70 mm. Annual average ETa estimates for the period 2003–2013 vary (Fig. 4.7.; Akhtar, 2017). ETa levels were the lowest during 2003–2004 within the range of 420–530 mm, mainly because of the severe drought from 1998–2004 (Fig. 4.7.). They were highest (560–580 mm) during the 2010–2013 period when this region experienced back-to-back floods. The drought period of 1998–2004 affected most of the region and had severe impacts, including the extinction of animals and plants, low crop productivity, and human migration.

The cryosphere

The cryosphere is crucial for determining and avoiding negative repercussions for the water resources in this region. Studies report that the glacier-covered area in the Kabul river basin is around 2.3–3.0% of the total basin area and during the winter most of the basin (on average

Fig. 4.4. Flow discharge measurement, Warsak Gravity Canal, Kabul river, Pakistan. (From: Kashif Hussain, IWMI-Pakistan)

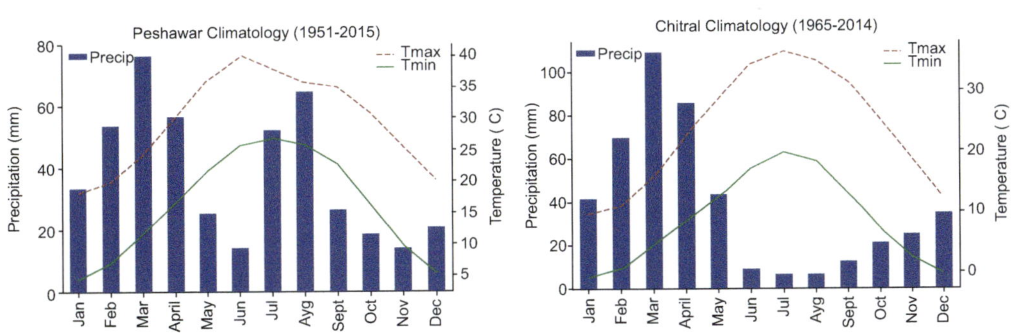

Fig. 4.5. Monthly variation in temperature and precipitation at Peshawar and Chitral in the Kabul river basin.

about 66%) is covered with snow (Masood *et al*., 2018; Taraky, 2021). Several studies have been conducted to determine how glacier and snow-covered areas vary spatially and temporally and contribute to runoff (Immerzeel and Bierkens, 2012). For example, Masood *et al*. (2018) explored aerial changes in the basin cryosphere from 2001 to 2016. They found that, on average, the inter-annual fluctuation of the total snow-covered area was within a range of 10–66%, with the lowest minimum snow-covered area of about 8% of the catchment area in a year. The highest maximum snow-covered area was around 77% in 2005. Fig. 4.8. depicts the monthly variability in the

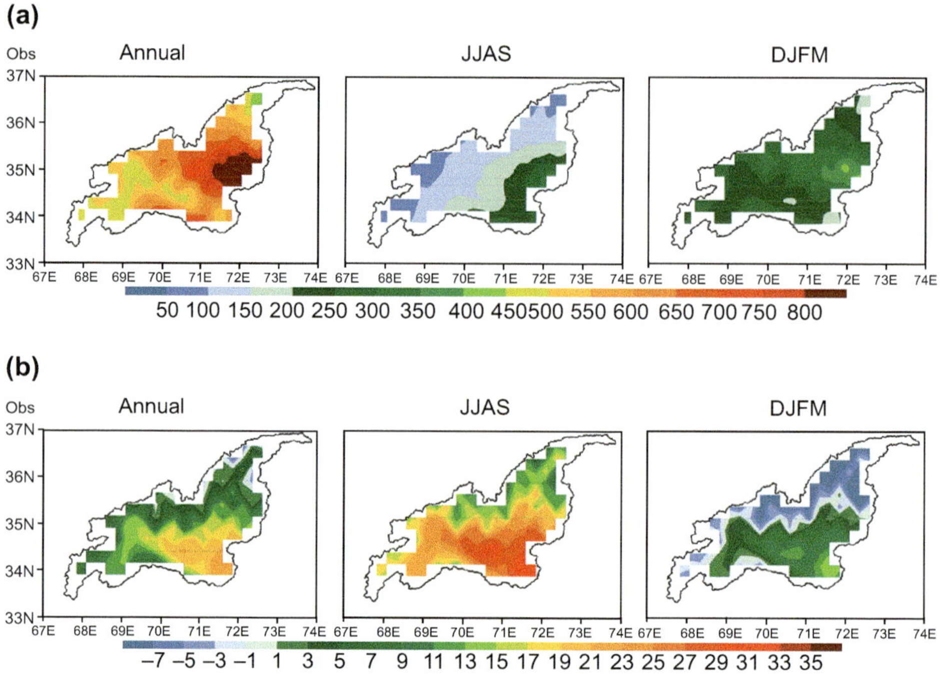

Fig. 4.6. (a) Mean precipitation (mm) 1975–2005; (b) Mean temperature (°C) 1975–2005. (Adapted from Bokhari *et al.*, 2018)

snow-covered area from 2001 to 2012. The snow cover begins to increase in August and continues to increase until February then declines from February to August. Because of the elevated temperatures and liquid precipitation, the minimum snow cover is seen from July to September, while the maximum is from December to February when the basin receives westerly precipitation, mainly snow (Masood *et al.*, 2018).

Snow cover trends. In terms of trends in snow cover in different elevation zones, analysis shows there is a significant increasing trend (Mann-Kendall Tau value in the range of 0.02–0.07) in snow-covered areas in the low-elevation zones (between 2000 and 2500 masl) and a decreasing trend in the higher-elevation zones (Masood *et al.*, 2018). The assessment of glacier mass balance in the Kabul river basin has not received much attention to date. A rare estimate suggests glaciers in the northern part of the basin have a positive mass balance, which may have significant implications for river flows and water availability (Mehmood *et al.*, 2021).

Kurram river basin

The Kurram river basin is relatively small compared to the Kabul basin (Fig. 4.9; Fig. 4.10). Meteorological records for this basin are available from two meteorological stations (i.e. Bannu and Parachinar, Table 4.2). The Kurram and the Gomal basins have extremely low levels of climate monitoring which requires immediate attention, otherwise these two basins will remain deprived of a comprehensive climate analysis, which is highly desirable considering the rapid changes in the regional climate and associated impacts.

Precipitation in the Kurram contains two peaks: March and July–August (Fig. 4.11.). Temperature minimum and maximum follow similar patterns, peaking in June and decreasing

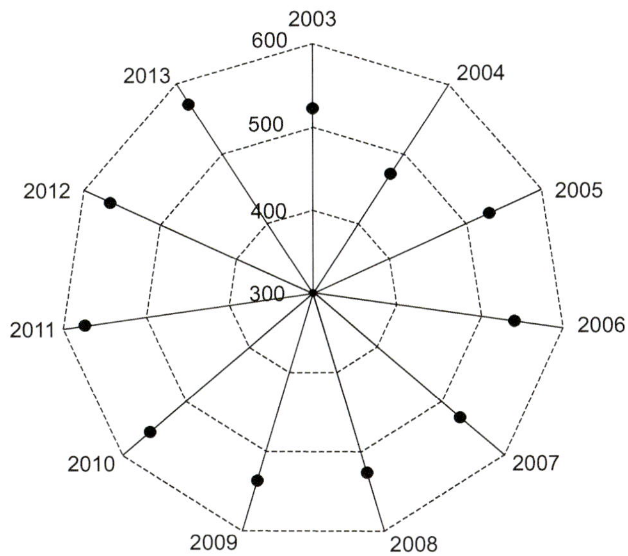

Fig. 4.7. Annual average ETa (mm) of the Kabul River Basin 2003–2013 (Source: Modified from Akhtar, 2017).

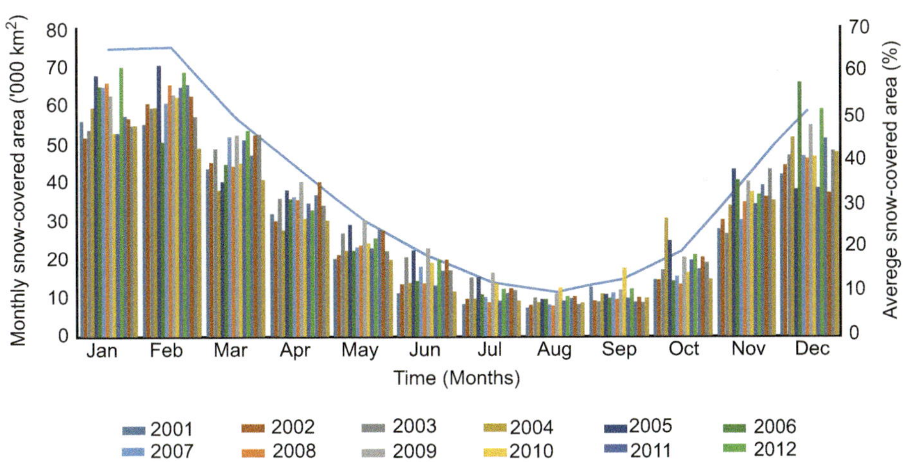

Fig. 4.8. Monthly variability of snow cover for the Kabul river basin between 2001 and 2012. (Adapted from Masood *et al.*, 2018)

steadily through December–January. These trends are consistent with the other two basins or the broader region. Closer scrutiny nonetheless reveals minor deviations.

The most up-to-date estimates provided by a recent study reported that Bannu has a long-term (1971–2018) mean annual rainfall of 362 mm, a mean annual potential evapotranspiration (PET) of 1691 mm, and an aridity index of 0.21 (semi-arid). Parachinar has a long-term mean annual rainfall of 987 mm, a mean annual PET of 1274 mm, and an aridity index of 0.78 (humid). There is an 18.27 mm/year increase in Parachinar precipitation, which is statistically significant. There

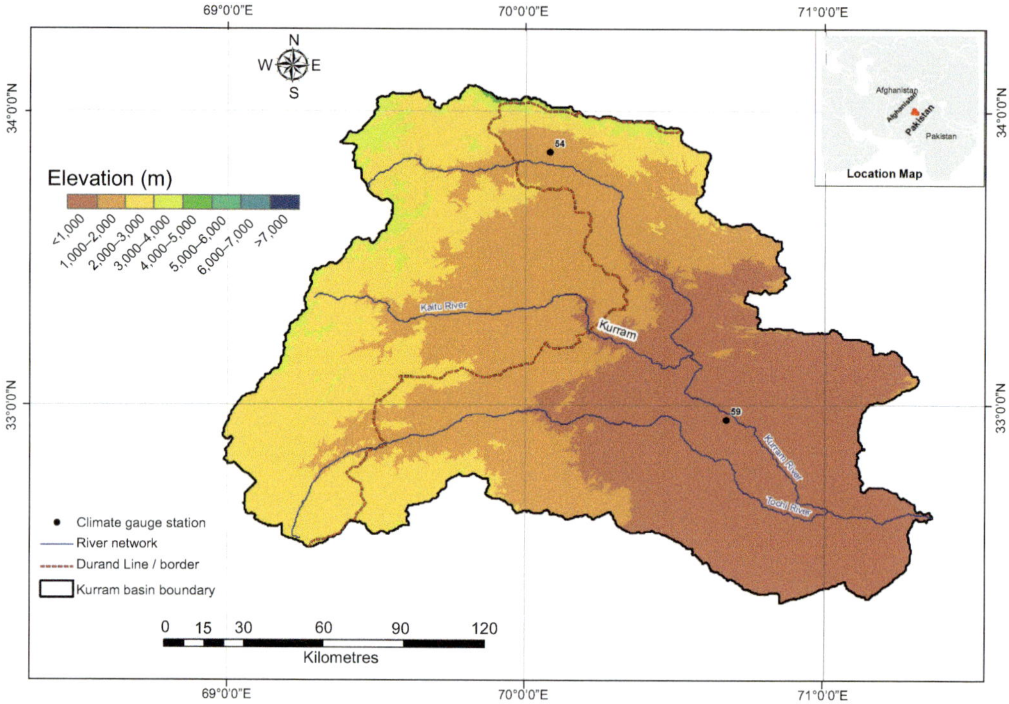

Fig. 4.9. Climate stations in the Kurram river basin.

Fig. 4.10. Kurram river, Pakistan. (From: Amjad Jamal, IWIMI-Pakistan)

Table 4.2. Climate stations in the Kurram river basin in Pakistan.

ID	Station	Parameter	Latitude	Longitude	Elevation (masl)
54	Parachinar	T, P	33.86	70.08	1725
59	Bannu	T, P	32.95	70.67	406

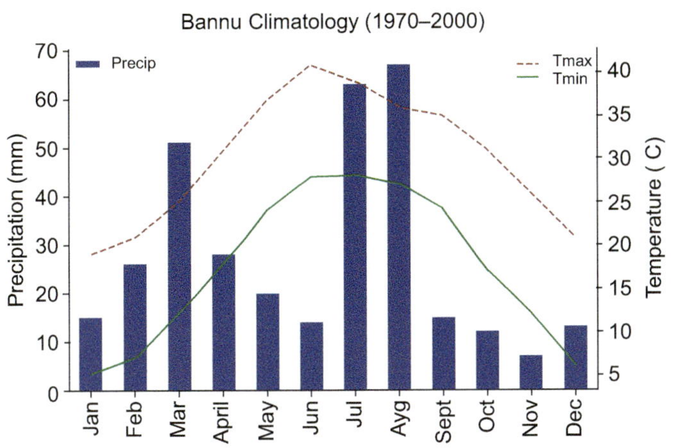

Fig. 4.11. Monthly variation in temperature and precipitation at Bannu in the Kurram river basin.

was a non-significant increasing trend detected for Bannu rainfall on a monthly (a majority of the months) and an annual basis. A significant increasing trend in the 12-month Standardized Precipitation Evaporation Index (SPEI) values for both Bannu and Parachinar was found, which is an indicator of an increasing trend for wet conditions in this area (Rahman *et al.*, 2021). Analysis of long-term data on temperature (similar to rainfall and PET) for this basin is not available as yet.

Gomal river basin

Among the three shared basins, the Gomal is the smallest in terms of watershed area and average annual river flow (Fig. 4.12; Fig. 4.13) (Rahman *et al.*, 2018; Bhatti *et al.*, 2021). There are two meteorological stations available that represent the climate in this basin (Table 4.3). A recent study by Rahman *et al.* (2021) provided some key climate statistics from at least one meteorological station. This study reported that the long-term (1971–2018) mean annual rainfall was 317 mm, the mean annual PET was 1748 mm, and the aridity index at this station was 0.18 (arid).

Precipitation in the Gomal contains two peaks: March and August (Fig. 4.14.). Temperature min and max follow similar patterns, although the maximum temperature peaks a bit earlier (May–June) than the min temperature (July–August) peaking. These trends are fairly consistent trends with the other two basins, but minor deviations are apparent.

In the case of annual rainfall, an increase of 21% in total volume was detected during the first decade of this century when compared to the average of the last 30 years. Seasonally, rainfall is projected to increase significantly in all seasons except the spring. The largest projected increase has been estimated for the autumn, which suggests that the duration of monsoons may extend towards September (Hussain *et al.*, 2013).

Climate change projections for the three basins

The Hindukush-Karakoram-Himalaya (HKH) and the downstream region have been reported as climate change hotspots (Lutz *et al.*, 2014, Lutz

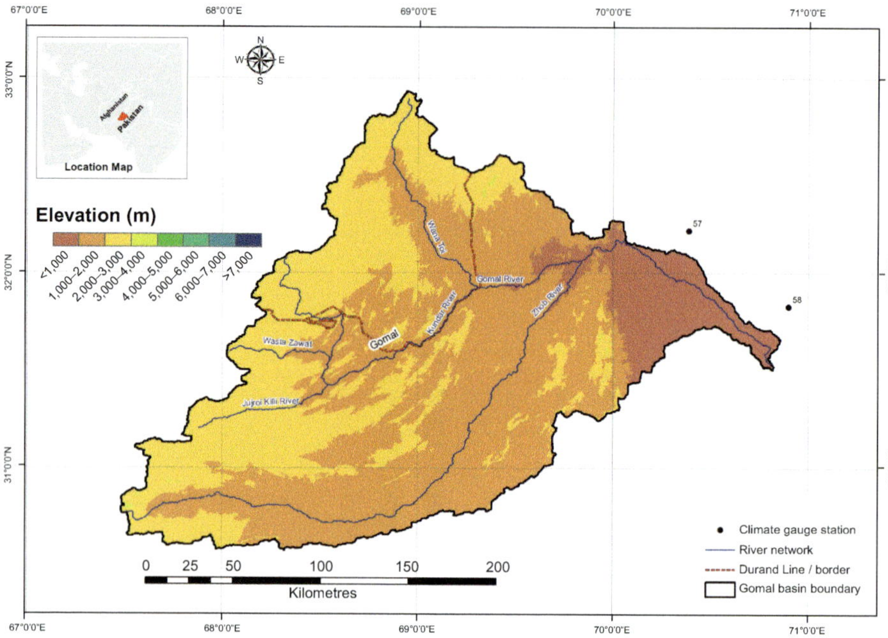

Fig. 4.12. Climate stations near the Gomal river basin.

Fig. 4.13. Gomal Zam dam, Gomal river basin, Pakistan. (From: Amjad Jamal-IWMI)

et al., 2016; Azmat *et al.*, 2018). The increase in average temperature in the HKH region (including Afghanistan and Pakistan) towards the end of the century is expected to be higher than that of the global average temperature (Iqbal *et al.*, 2018).

Table 4.3. Climate stations, Gomal river basin.

ID	Station	Parameter	Length of record	Latitude	Longitude	Elevation (masl)
57	Tank	T, P	1973–2017	32.22	70.39	481
58	Dera Ismail Khan	T, P	1971–2018	32.22	70.39	171.2

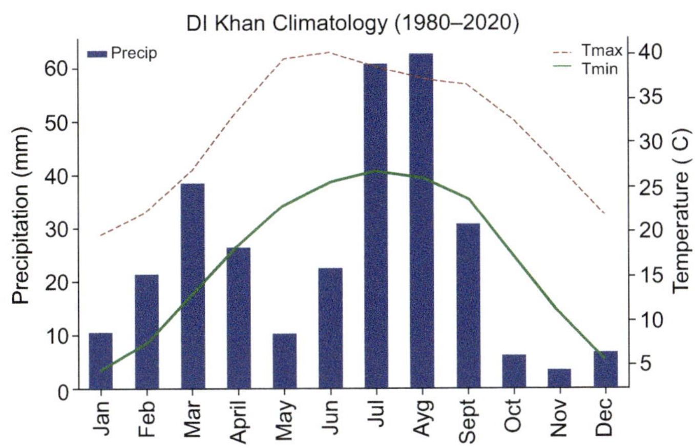

Fig. 4.14. Monthly variation in temperature and precipitation at DI Khan in the Gomal river basin.

Several studies have attempted to predict climate change in the Kabul river basin. Future climate change will not be uniform owing to topographic gradients and more than one climate prevailing in different parts of the basin. However, there seems to be a consensus regarding a significant increase in temperatures across the basin and considerable uncertainties about the future rates of precipitation (e.g Sidiqi *et al.*, 2018; Hashmi *et al.*, 2020; Azizi and Asaoka, 2020).

Table 4.4 summarizes the main findings of the most prominent research studies on climate change projections in the Kabul river basin in terms of estimated future change in the most important climate variables.

The main findings of the studies presented in Table 4.4 are

- Temperature is rising and will continue to rise throughout the 21st century.
- By the end of the 21st century, temperatures may rise within a range of 3–4°C under RCP 4.5 (mitigation scenario) and 5–6°C under RCP 8.5 (business as usual scenario).
- There is a high level of uncertainty about future changes in precipitation in the basin; however, there is a general tendency towards decreasing precipitation in the autumn and winter.

Some studies have reported that although there is a spatial variation in agricultural and meteorological droughts in the Kabul, Kurram and Gomal river basins, severe-to-moderate droughts showed a decreasing temporal trend in the early part of this century (i.e. 2000–2018) for both types of droughts, and a similar trend is projected to continue in the 2020s and 2030s (Baig *et al.*, 2020).

No study was found that provides information relating to climate change projections for the Kurram river basin and there is only one attempt by Hussain *et al.* (2013), who found near-term (2030s) climate change estimates on a decadal basis for the Dera Ismail Khan station (representing the Gomal river basin) project an average decadal increase of 0.4°C in annual mean temperature, a decrease in winter temperatures by 0.3°C and a 0.7°C increase in spring temperatures, causing early summers. No significant trend was found for summer and autumn temperatures (Hussain *et al.*, 2013).

Table 4.4. Summary of climate projections in the three basins.

Reference	Variable	RCP 4.5 Scenario	RCP 8.5 Scenario	Future periods
Wi *et al.*, 2015	T	2.2	2.8	2050s
	P	-	-	No clear trend
	PET (mm)	+100	+150	2050s
Ghulami, 2017	Tmax	1.7–4.5	1.7–6.3	Early to late 21st century
	Tmin	1.5–3.8	1.4–6	
	P (%)	-19–-6	-18–3	
Bokhari *et al.*, 2018	T	1.5–1.8 2.7–3.0 3.2–3.7	1.6–2.0 3.5–4.0 Up to 6.4	2010–2039 2040–2069 2070–2100
	P	20– -40 -35– -40 -30– -45	-15– -40 -25– -35 -15– -35	2010–2039 2040–2069 2070–2100
Sidiqi *et al.*, 2018	Tmax	3.9	6.2	2080s
	Tmin	3.4	5.7	2080s
	Tav	1.7 3 3.7	2 3.9 6	2020s 2050s 2080s
Hashmi *et al.*, 2020	T	3.1 4.5	3.2 5.0	2020s 2050s
	P	1.3 1.4	1.4 1.7	2020s 2050s
Ghulami *et al.*, 2020	Tmax Tmin	+1.05–+3.13 +0.63–+2.94	+0.85 –+5.83 +0.63 –+5.23	Early–late 21st century
	P	Decreasing trend in winter; increasing trend in summer	Decreasing trend in winter; increasing trend in summer	
Khatiwada *et al.*, 2021	T	3	5	By end of 21st century
	P (%)	4	12	
Khatiwada *et al.*, 2021 (for the Panjshir sub-basin)	T	1.9–3.1 2.5 – 4.3	2.2–4.4 4.9–7.8	Mid-century End-century
	P (%)	-17–4 -26–13	-34–23 -40–28	Mid-century End-century
	Snowfall amount	Decrease in summer and autumn		Mid–end 21st century

T, temperature;P, precipitation;PET, potential evapotranspiration

Climate Change Governance: An Opportunity for Collaboration?

Both Afghanistan and Pakistan are among the countries most vulnerable to the impacts of climate change. In 2021, the ND-GAIN Index ranked Afghanistan at 175 and Pakistan at 151 out of 181 countries (University of Notre Dame, 2021). Based on this index, the lower score indicates a higher vulnerability to climate change.

Both Afghanistan and Pakistan face higher warming rates compared to the global average. Changes in precipitation and runoff regimes in both countries are highly uncertain and drought and flood incidents are likely to increase. The increases in the annual minimum and maximum temperatures in both Afghanistan and Pakistan are predicted to be greater than the increase in average temperature, which will increase pressure on human health, livelihoods and ecosystems (ADB (Asian Development Bank) and World Bank, 2021).

In Afghanistan, the Climate Change Directorate has been established based on requirements in the 2010 National Environmental Protection Agency Framework. This Directorate works on climate change and its impacts throughout the country. The Directorate is organized into three sub-directorates:

- Climate Change Adaption Sub-Directorate
- Climate Change Mitigation Sub-Directorate
- Climate Change Finance Sub-Directorate

In terms of its mandate, the work of the Climate Change Directorate is based on environmental law and Afghanistan's commitments to UNFCCC objectives. The Directorate is responsible for addressing climate change in terms of both adaptation and mitigation and has the following mandate:

- to establish effective co-ordination in all aspects of climate change between line ministries, the private sector, national NGOs and multilateral implementing entities;
- to monitor and evaluate the implementation of climate change policies, strategies, and long- and mid-term planning from our stakeholders;
- to mainstream climate change issues into national policies and development planning;
- to raise public awareness related to adverse impacts of climate change at the community level, to build capacities against the impacts of climate change;
- to monitor climate change projects and remain involved in all stages of the project with the bilateral and multilateral implementing entities during the implementation of these projects;
- to identify the causes of greenhouse gas (GHG) emissions at the national level;
- to avoid anthropogenic interventions and stabilize the concentration of GHGs in the atmosphere;
- to reduce GHG emissions and increase carbon sinks; and
- to assess vulnerability, build adaptive capacity and increase resilience to sustain the ecosystem and decrease threats from expected climatic risks.

Pakistan's policy response to the challenges of climate change has already been incorporated into its Strategic Vision 2025, its commitments to the Sustainable Development Goals, and the objectives of the United Nations Framework Convention on Climate Change (UNFCCC). Adoption of the National Climate Change Policy and National Disaster Risk Reduction Policy 2012 provides a comprehensive framework for policy goals and actions towards mainstreaming climate change. It specifically addresses the economically and socially vulnerable sectors of the economy. A follow-up to these policies was the launch of the Framework for Implementation of the Climate Change Policy 2014–2030, which outlines the vulnerabilities of various sectors to climate change and identifies appropriate adaptation and mitigation actions (Climate Change Division, 2013).

Although climate change became a provincial matter after the 18th amendment of the Constitution of Pakistan, some activities and responsibilities relating to climate change concerns are co-ordinated by the Ministry of Climate Change at the national level (e.g. climate change assessments and reporting under UNFCCC) with corresponding support from subnational governments.

The Ministry of Climate Change works in close collaboration with federal government departments, research institutions, universities and the private sector and is also responsible for supervising and controlling several attached departments, autonomous bodies and implementation agencies such as:

- the Global Change Impact Studies Centre;
- the Pakistan Environmental Protection Agency;
- the Zoological Survey Department of Pakistan; and
- the Islamabad Wildlife Management Board.

There is, as yet, no treaty, agreement or joint plan for transboundary basin governance and climate change adaptation between

Afghanistan and Pakistan. However, given the climate threats the three basins are facing, there is an opportunity for sustained co-operation between the two countries. The following are four policy recommendations that could be pursued:

- A joint campaign by the meteorological agencies of the two countries to install climate and river flow gauging sites, especially on the Afghanistan side, will eventually have long-term data available for reliable climate analysis. This would help improve hydrological monitoring and climate change impact assessment research studies in the basin.
- Joint watershed management practices, including tree plantations, to reduce soil erosion and flood risk, and flood plain management (restoring and creating wetlands) to make beneficial use of flood flows.
- A formal mechanism for co-operation and co-ordination among the national weather and flood forecasting agencies in both countries through enhanced knowledge and data sharing and mutual capacity building.
- Jointly run flood early warning systems for saving, as much as possible, precious lives and property in both countries. In future, joint modelling studies for this basin should be carried out by researchers in both countries and a multiple-model hydrological modelling approach should be used to quantify uncertainties in future flow projections.

Conclusions

This chapter has crystallized four main messages. First, the most concerning finding is the limited duration and quality of the available climatic data, particularly in high-elevation regions of the Kabul and the entire Gomal and Kurram river basins. Unless immediate steps are taken jointly by Afghanistan and Pakistan to enhance hydrometeorological observations, it will be extremely difficult, if not impossible, to conduct climate change assessments and formulate appropriate mitigation and life-saving adaptation strategies for vulnerable communities.

Second, climate change is becoming more apparent. The current climate variability and trends indicated by available data analysis studies suggest that temperatures in the three basins have been rising consistently. While future climate projections indicate that by the end of the 21st century temperatures may rise within a range of 3–4°C under RCP 4.5 (mitigation scenario) and 5–6°C under RCP 8.5 (business as usual scenario), there is considerable uncertainty associated with the current and future behaviour of precipitation in the three basins, mainly due to temporal and spatial gaps in the available climate data and the short lengths of the available records.

Third, climate change enhances uncertainty which constrains the formulation of policy responses. Consistently rising temperatures have major implications for snow, glacier coverage and water availability for the major socio-economic sectors. The uncertainty associated with the behaviour of precipitation in the three basins is a major challenge to ensuring sustainably fulfilling water requirements for various sectors, especially agriculture. Due to a rise in temperature, there is evidence of faster snowmelt, which has caused glacial lake outburst floods and snow avalanches in the mountainous region of the Kabul river basin and riverine floods in both the Afghan and Pakistani portions of the Kabul, Kurram and Gomal river basins. Such events are expected to happen more frequently and with greater intensity.

Fourth, the adverse effects of floods may increase. Increasing trends in rainfall, as reported by some studies, may further increase the possibility of flash-flooding events in both the high- and low-elevation parts of the three basins. Climate change may also worsen droughts in the areas which either already receive low amounts of rainfall or depend heavily on snow- and glacier-melt water. Potential solutions to avoid the negative repercussions of climate change implications include:

- urgently enhancing the climate monitoring network, especially in the high-elevation parts of the Kabul river basin and the entire Kurram and Gomal river basins;

- conducting joint studies on the impacts of changing climate, which can be multi-sectoral as well;
- mitigating hydro-meteorological disasters by jointly run early-warning systems and efficient data sharing;
- introducing water-saving technologies in agriculture to achieve climate-smart agriculture; and
- introducing low-delta, drought-tolerant crop varieties.

Note

[1] RCP- Representative Concentration Pathway

Acknowledgement

The authors greatly appreciate the contribution of Mr Haris Mushtaq in the preparation of this chapter.

References

ADB (Asian Development Bank) and World Bank (2021) Climate risk country profile: Afghanistan. Available at: https://www.adb.org/sites/default/files/publication/660566/climate-risk-country-profile-afghanistan.pdf (accessed 23 July 2022).

Akhtar, F. (2017) *Water availability and demand analysis in the Kabul River Basin, Afghanistan*. Doctoral dissertation, University of Bonn, Germany.

Azizi, A.H. and Asaoka, Y. (2020) Assessment of the impact of climate change on snow distribution and river flows in a snow-dominated mountainous watershed in the western Hindukush–Himalaya, Afghanistan. *Hydrology* 7(4), 74. DOI: 10.3390/hydrology7040074.

Azmat, M., Qamar, M.U., Huggel, C. and Hussain, E. (2018) Future climate and cryosphere impacts on the hydrology of a scarcely gauged catchment on the Jhelum river basin, Northern Pakistan. *The Science of the Total Environment* 639, 961–976. DOI: 10.1016/j.scitotenv.2018.05.206.

Baig, M.H.A., Abid, M., Khan, M.R., Jiao, W., Amin, M. *et al*. (2020) Assessing meteorological and agricultural drought in Chitral Kabul river basin using multiple drought indices. *Remote Sensing* 12(9), 1417. DOI: 10.3390/rs12091417.

Bhatti, M.T., Ashraf, M. and Anwar, A.A. (2021) Soil erosion and sediment load management strategies for sustainable irrigation in arid regions. *Sustainability* 13(6), 3547. DOI: 10.3390/su13063547.

Bokhari, S.A.A., Ahmad, B., Ali, J., Ahmad, S., Mushtaq, H. *et al*. (2018) Future climate change projections of the Kabul river basin using a multi-model ensemble of high-resolution statistically downscaled data. *Earth Systems and Environment* 2(3), 477–497. DOI: 10.1007/s41748-018-0061-y.

Casale, F., Bombelli, G.M., Monti, R. and Bocchiola, D. (2020) Hydropower potential in the Kabul River under climate change scenarios in the XXI century. *Theoretical and Applied Climatology* 139(3–4), 1415–1434. DOI: 10.1007/s00704-019-03052-y.

Climate Change Division (2013) *Framework for Implementation of Climate Change Policy Government of Pakistan (2014–2030)*. Available at: http://www.gcisc.org.pk/Framework%20for%20Implementation%20of%20CC%20Policy.pdf (accessed 5 October 2022).

Ghulami, M. (2017) Assessment of climate change impacts on water resources and agriculture in data scarce Kabul basin, Afghanistan. Doctoral dissertation, Université Côte d'Azur and Asian Institute of Technology.

Ghulami, M., Gourbesville, P. and Audra, P. (2020) Assessing future water availability under a changing climate in Kabul Basin. In: Gourbesville, P. and Caignaert, G. (eds) *Advances in Hydroinformatics*.

Springer Singapore, pp. 647–657. Available at: https://doi.org/10.1007/978-981-15-5436-0_50 (accessed 19 September 2022).

Guan, Y., Zhang, X., Zheng, F. and Wang, B. (2015) Trends and variability of daily temperature extremes during 1960–2012 in the Yangtze River Basin, China. *Global and Planetary Change* 124, 79–94. DOI: 10.1016/j.gloplacha.2014.11.008.

Gurung, T.R., Acharya, A., Dildar, M. and Joshi, S.P. (2020) Preliminary hydrometeorological data analysis of the Kabul River basin, Afghanistan: need for a long-term glacier monitoring programme. Available at: https://agris.fao.org/agris-search/search.do?recordID=QZ2022000043 (accessed 23 July 2022).

Hashmi, M.Z. ur R., Masood, A., Mushtaq, H., Bukhari, S.A.A., Ahmad, B. *et al.* (2020) Exploring climate change impacts during first half of the 21st century on flow regime of the transboundary Kabul River in the Hindukush region. *Journal of Water and Climate Change* 11(4), 1521–1538. DOI: 10.2166/wcc.2019.094.

Hassanyar, M.H. and Tsutsumi, J. (2017) Multi-model ensemble climate change projection for Kunduz River Basin, Afghanistan under representative concentration pathways. *Modern Environmental Science and Engineering* 03(05), 291–301. DOI: 10.15341/mese(2333-2581)/05.03.2017/001.

Hussain, S.S., Hussain, S., Hanif, M., Nizami, A., Pakistan, I. *et al.* (2013) Climate change scenarios and possible adaptation measures. Districts Chitral and Dera Ismail Khan. Intercooperation Pakistan, Islamabad.

Immerzeel, W.W. and Bierkens, M.F.P. (2012) Asia's water balance. *Nature Geoscience* 5(12), 841–842. DOI: 10.1038/ngeo1643.

Iqbal, M., Dahri, Z., Querner, E., Khan, A. and Hofstra, N. (2018) Impact of climate change on flood frequency and intensity in the Kabul river basin. *Geosciences* 8(4), 114. DOI: 10.3390/geosciences8040114.

Khatiwada, K.R., Pradhananga, S. and Nepal, S. (2021) Potential implications of future climate change and extremes on health, agriculture, water, and cryosphere sectors in the Kabul river basin. DOI: 10.21203/rs.3.rs-645615/v1.

Lutz, A.F., Immerzeel, W.W., Shrestha, A.B. and Bierkens, M.F.P. (2014) Consistent increase in High Asia's runoff due to increasing glacier melt and precipitation. *National Climate Change* 4, 587–592. DOI: 10.1038/NCLIMATE2237.

Lutz, A.F., Immerzeel, W.W., Kraaijenbrink, P.D.A., Shrestha, A.B. and Bierkens, M.F. (2016) Climate change impacts on the upper Indus hydrology: sources, shifts and extremes. *PloS One* 9(11), e0165630. DOI: 10.1371/journal.pone.0165630.

Masood, A., Hashmi, M.Z. ur R. and Mushtaq, H. (2018) Spatio-temporal analysis of early twenty-first century area changes in the Kabul river basin cryosphere. *Earth Systems and Environment* 2(3), 563–571. DOI: 10.1007/s41748-018-0066-6.

Mehmood, A., Jia, S., Lv, A., Zhu, W., Mahmood, R. *et al.* (2021) Detection of spatial shift in flood regime of the Kabul River Basin in Pakistan, causes, challenges, and opportunities. *Water* 13(9), 1276. DOI: 10.3390/w13091276.

Rahman, G., Atta-ur-Rahman, S. and Dawood, M. (2018) Spatial and temporal variation of rainfall and drought in Khyber Pakhtunkhwa Province of Pakistan during 1971–2015. *Arabian Journal of Geosciences* 11(3). DOI: 10.1007/s12517-018-3396-7.

Rahman, G., Rahman, A.-U., Ullah, S., Dawood, M., Moazzam, M.F.U. *et al.* (2021) Spatio-temporal characteristics of meteorological drought in Khyber Pakhtunkhwa, Pakistan. *PLOS ONE* 16(4), e0249718. DOI: 10.1371/journal.pone.0249718.

Salehi, S., Dehghani, M., Mortazavi, S.M. and Singh, V.P. (2020) Trend analysis and change point detection of seasonal and annual precipitation in Iran. *International Journal of Climatology* 40(1), 308–323. DOI: 10.1002/joc.6211.

Shroder, J. and Ahmadzai, S.J. (2016) *Transboundary Water Resources in Afghanistan – Climate Change and Land-Use Implications*. Elsevier, Amsterdam.

Sidiqi, M., Shrestha, S. and Ninsawat, S. (2018) Projection of climate change scenarios in the Kabul River Basin, Afghanistan. *Current Science* 114(06), 1304. DOI: 10.18520/cs/v114/i06/1304-1310.

Taraky, Y.M. (2021) Transboundary water resources in the context of developmental goals and climate change, the case of the Kabul river basin. Doctoral dissertation, University of Guelph, Ontario, Canada.

University of Notre Dame (2021) Notre Dame Global adaptation initiative. Country Index. Available at: https://gain.nd.edu/our-work/country-index/ (accessed 23 July 2022).

Wang, X., Tolksdorf, V., Otto, M. and Scherer, D. (2021) WRF-based dynamical downscaling of ERA5 reanalysis data for High Mountain Asia: towards a new version of the High Asia Refined analysis. *International Journal of Climatology* 41(1), 743–762. DOI: 10.1002/joc.6686.

Wi, S., Yang, Y.C.E., Steinschneider, S., Khalil, A. and Brown, C.M. (2015) Calibration approaches for distributed hydrologic models in poorly gaged basins: implication for streamflow projections under climate change. *Hydrology and Earth System Sciences* 19(2), 857–876. DOI: 10.5194/hess-19-857-2015.

Yadav, R.K., Yoo, J.H., Kucharski, F. and Abid, M.A. (2010) Why is ENSO influencing northwest India winter precipitation in recent decades? *Journal of Climate* 23(8), 1979–1993. DOI: 10.1175/2009JCLI3202.1.

Yaseen, M., Ahmad, I., Guo, J., Azam, M.I. and Latif, Y. (2020) Spatiotemporal variability in the hydrometeorological time-series over upper Indus River Basin of Pakistan. *Advances in Meteorology* 2020, 1–18. DOI: 10.1155/2020/5852760.

5 Surface Water

Hassaan Furqan Khan and Arif Anwar

Chapter Overview

Surface waters of the transboundary Kabul, Kurram and Gomal basins play a key role in shaping cultural, social and economic lives in Afghanistan and Pakistan. Discussing each of the basins separately, the chapter assesses hydrology and water resources. iIt includes a description of the existing and planned water infrastructure, as well as a review of relevant surface water quality issues. A major portion of this chapter is devoted to the Kabul river basin, not only because of its significantly larger size and much greater surface water flows, but also because of the limited availability of relevant hydrologic datasets and published scientific studies in the Kurram and Gomal basins, where grey literature has been heavily relied upon.

Fig. 5.1. Gomal river downstream of Gomal Zam Dam, Pakistan. (From: Amjad Jamal, IWMI)

DOI: 10.1079/9781800622371.0005

Introduction

Surface waters comprise the main source of water in the transboundary Kabul, Kurram and Gomal basins. Surface water is essential for large-scale water uses such as agriculture in both countries and is also critical to current and future hydropower generation in the basins (Akhtar and Iqbal, 2017). Surface water is nonetheless variable in space and time, which drives the need to understand hydrology and water resource use so that benefits from water-dependent sectors are maximized. While understanding hydrology and water resources in all basins shared by Afghanistan and Pakistan is important, as outlined in Chapter 1, the Kabul, Kurram and Gomal can be considered most important given their relative size, development level and water uses.

The chapter considers existing hydrology and potential future hydrological changes, as well as water quality, surface water storage and hydropower development. This is believed to be one of the first attempts to examine the hydrology of these three basins by incorporating data from both countries. Following a brief description of data acquisition, the chapter devotes sections to the Kabul, Kurram and Gomal river basins. A major portion of this chapter is devoted to the Kabul river basin, not because of its significantly larger size and much greater surface water flows, but because of the greater availability of relevant hydrological datasets and published scientific studies. A smaller portion of the chapter is devoted to the Kurram and Gomal for which future hydrological changes could not be sufficiently covered.

Data Compilation and Processing

Data compilation

Access to streamflow time-series data across the Kabul, Kurram and Gomal basins was challenging. Streamflow data for gauges situated in Pakistan were collected primarily from the Surface Water Hydrology Project maintained by the Water and Power Development Authority (WAPDA). In some cases, the Surface Water Hydrology Project data was obtained directly from WAPDA while for a few gauges (in Kurram and Gomal basins), the data was extracted from feasibility reports of the Planning Commission. Data from streamflow gauges in Afghanistan were obtained primarily from the Global Runoff Data Center (GRDC) and the United States Geologic Survey (USGS). Due to the prevailing political situation, we were unable to obtain streamflow data from the Afghan Ministry for Energy and Water.

Data harmonization

For this chapter, streamflow data was collected from a variety of sources. Each source employs different data storing and formatting practices. Significant effort was expended in collating all the various datasets into a uniform format to allow for subsequent analyses.

Analyses in a context of limited data

Monitoring surface water flows in the Kabul basin in Pakistan is relatively extensive, with over 30 flow gauges recording observations over at least 30 years (Mehmood *et al.*, 2021). Monitoring in Pakistani portions of the Kurram and Gomal river basins is less extensive, but data points do exist. In contrast, there are significantly fewer gauging stations in Afghanistan, with time series data at most of these gauges displaying many gaps owing to the political turmoil over the past 45 years. Ultimately, an effort was made to mine limited data to provide a reasonable characterization of conditions. In the case of the Kabul, the selection of streamflow gauging stations used in the analysis was based on (i) data availability; and (ii) key confluence locations.

Kabul River Basin

Hydrology

Spread over 92,000 km^2, the Kabul river basin plays an integral role in the water resources of both Afghanistan and Pakistan. The basin's waters are critical to sustaining agriculture, which accounts for most of both land and water use. The Kabul river originates in the Paghman Mountains and flows eastward through Kabul and Jalalabad. The total average annual flows suggest that the upper Kabul generates only about 2.6% of the average annual flow of the basin (Iqbal *et al.*, 2018). In terms of flow contribution, the transboundary Kunar river emerges as the largest tributary to the Kabul river, followed by the Swat river. The Kunar originates in Pakistan, where it is called the Chitral river, and flows southward through Afghanistan joining the Kabul river near Jalalabad. It then continues into Pakistan's Khyber Pakhtunkhwa province (KPK) where it eventually joins the Indus river. Much of the annual discharge in the Kunar river flows into the Chitral catchment in Pakistan (Fig. 5.3) (Khattak *et al.*, 2015).

Upstream of the confluence with the Kunar river, the Kabul river registers annual average flows of 5.7 billion cubic metres (bcm) at Daronta, which are then augmented by another 15 bcm through the confluence of the Kunar river. At Dakah, the last gauging station on the Kabul river in Afghanistan, annual average flows are approximately 15.9 bcm, suggesting significant surface water use in Nangarhar province, mostly for agricultural purposes (Najmuddin *et al.*, 2021). At Nowshera, just upstream of the confluence of the Kabul river with the Indus, the average annual flows between 1960 and 2012 were approximately 23.8 bcm, making Kabul river the third biggest tributary to the Indus.

Table 5.1 outlines the extent of streamflow time series for each of the stream gauges shown in Fig. 5.3.

A review of average monthly flows at different gauge stations in the Kabul river basin suggests a pattern of flow typical of other rivers in the upper Indus basin (Fig. 5.4). Streamflow in the basin is highly seasonal, with snow and glacier melt in the spring and summer seasons the major contributing sources. Streamflow

Fig. 5.2. Automatic flow measurement on Warsak gravity canal, Kabul River in Pakistan. (From: Kashif Hussain, IWMI)

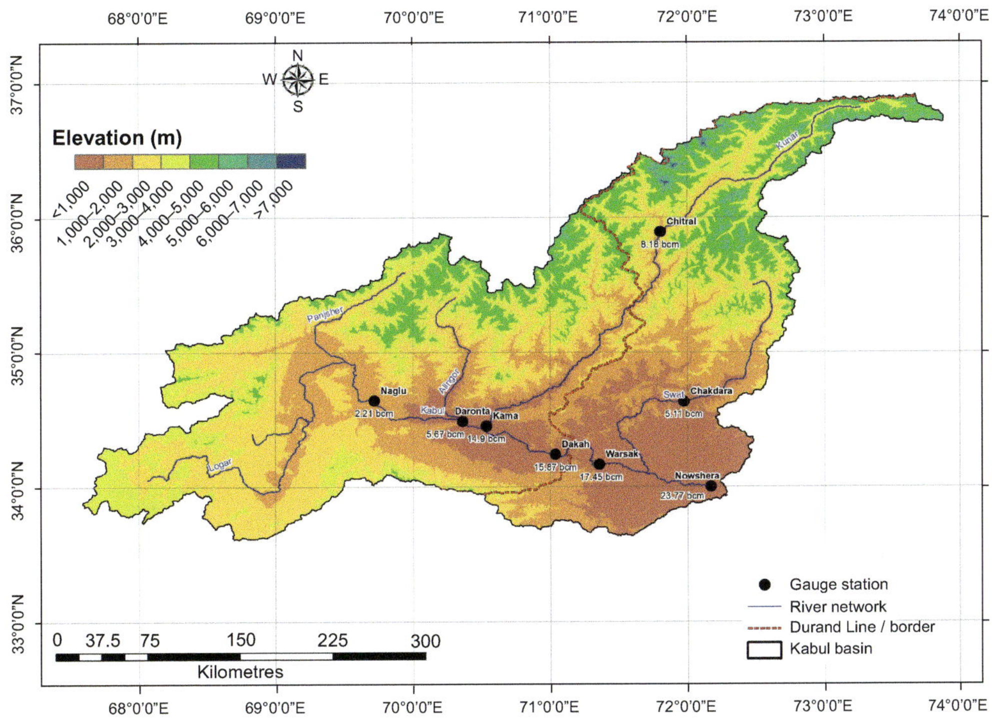

Fig. 5.3. Annual average flows (bcm) measured across the transboundary Kabul river basin. (Adapted from: Khattak *et al*., 2015)

Table 5.1. Streamflow gauging stations across the Kabul river basin used in this analysis. (Adapted from Surface Water Hydrology Project (WAPDA), United States Geologic Survey, Global Runoff Data Center)

Station	River	Data start	Data end
Naghlu	Kabul	September 2008	September 2013
Daronta	Kabul	October 1959	September 1964
Pul-e-Kama	Chitral/Kunar	January 1967	September 1979
Dakah	Kabul	April 2007	September 2014
Warsak	Kabul	February 1962	December 2005
Chitral	Chitral/Kunar	January 1964	December 2007
Chakdara	Swat	January 1961	December 2005
Nowshera	Kabul	January 1961	December 2019

peaks in all the rivers in either June or July. The rate of increase in flow during the spring varies across all rivers, with the Swat river showing the steepest rate of increase.

There is a notable variation in total annual streamflow measured at Nowshera from 1960 to 2020 (Fig. 5.5). Mean annual streamflow during this period was 23.8 bcm, ranging from a low of 13 bcm to a high of 43 bcm. The highest flows were observed during historic floods in 1991–1992. Previous analyses have found no evidence of a statistically significant temporal trend in the data (Akhtar and Iqbal, 2017). This total annual flow time series displays a notable

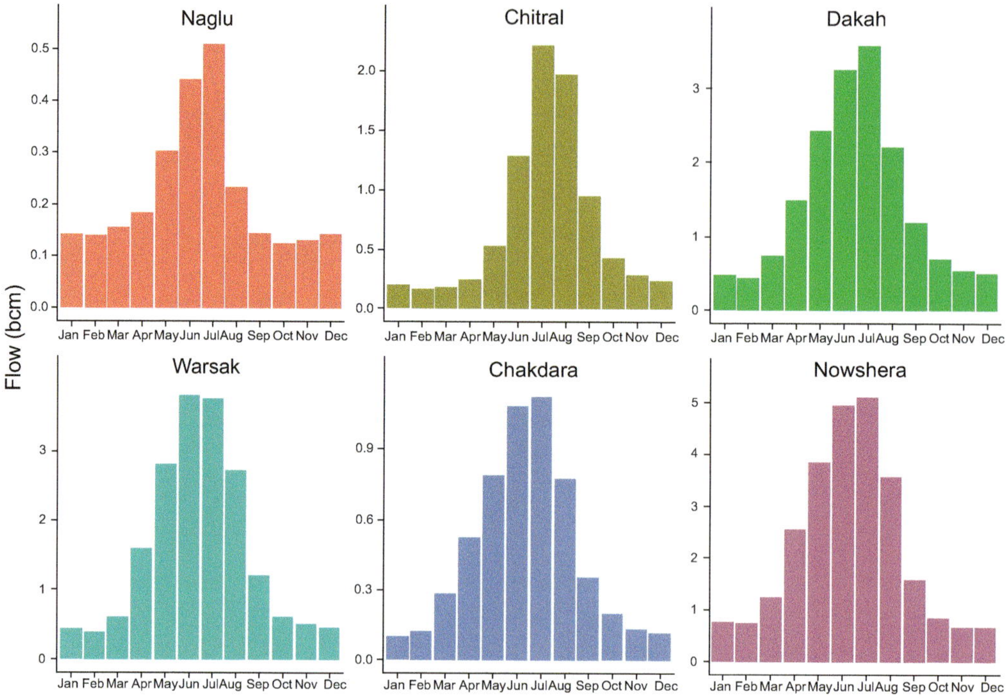

Fig. 5.4. Average monthly flows (bcm) across the Kabul river basin.

coefficient of variation (CV) of 0.258. The total annual flow for 2010 is unavailable in the time series data provided by WAPDA. This may be due to damage to the gauge from major flooding that year.

Flooding is an ever-present danger in the basin and has, over the years, been the primary hazard of concern. Flooding events in the region typically take place during the monsoon season (i.e. July through September), with the most significant floods generated by a combination of monsoon rains and snow and glacier melt. Both anecdotal and empirical evidence indicates concerning trends in average and extreme streamflow across the basin (Latif *et al.*, 2021). An extensive trend analysis of annual extremes (floods) across the gauges in the Kabul river basin in Pakistan found that the flood regime is non-stationary (Mehmood *et al.*, 2019). The Chitral and Swat rivers showed an increasing trend in both the frequency and magnitude of floods. Compared to the central and south-western parts of the Kabul basin, the rate of change in the flood regime was higher in the north and north-east (Hashmi *et al.*, 2020). Accounting for this non-stationarity in the design of planned water infrastructure is extremely important.

Future changes

Projected increases in future temperature may affect the winter snowpack in the Hindukush region which is a major source of regional streamflow. Studies indicate a decreasing trend in meteorological droughts in the northern parts of the Kabul river basin (e.g. Baig *et al.*, 2020). Given the high uncertainty in both the direction and magnitude of future changes in precipitation for the basin, the resulting impact on streamflow in the medium-to-long term remains an active area of research (Mehmood *et al.*, 2021).

Modelling streamflow in the Kabul river basin is challenging due to a lack of sufficient *in situ* measurement of hydroclimatic variables (especially in Afghanistan) and limited glacier-monitoring programmes. This dearth of information means high uncertainty in the

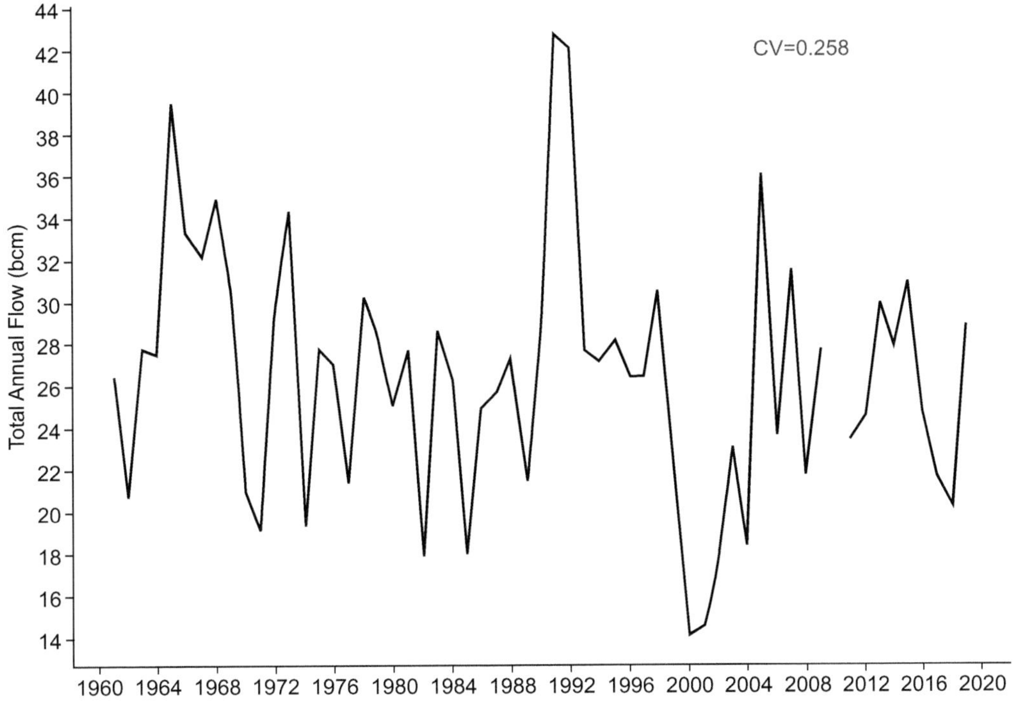

Fig. 5.5. Total annual streamflow measured at Nowshera, the most downstream gauge located on the Kabul river basin. Data for 2010 are missing. (Adapted from Surface Water Hydrology Project, WAPDA)

disaggregation of the baseflow and runoff components of streamflow as well as the identification of the snow and glacier melt contribution to flows (Bokhari *et al.*, 2018). This, in turn, affects confidence in the projection of flows in the Kabul river considering climate change. Nevertheless, physically and statistically based hydrologic models have been developed for various parts of the basin (Wi *et al.*, 2015; Aawar and Khare, 2020; Mehmood *et al.*, 2021). These modelling efforts foresee an increase in surface water flows in the short term due to increased snowmelt associated with higher temperatures in the region, with a decrease in flows in the long term with the receding of glaciers (Khan *et al.*, 2020; Taraky *et al.*, 2022).

Water quality

The waters of the Kabul river provide a range of ecosystem services including irrigation, effluent discharge, aquatic habitats, water for domestic use, and recreation (Akhtar and Iqbal, 2017). Since 2001, the doubling of the population and industrial sites has led to deteriorated water quality and lower crop yields, and periodic fish kills and skin diseases associated with poor water quality have been reported (Iqbal *et al.*, 2019).

The relatively limited number of empirical studies of water quality on the Kabul river (Akhtar and Iqbal, 2017) paints a picture of a severely stressed river facing several challenges. There are extremely low levels of dissolved oxygen coupled with high levels of chemical and biological oxygen demand (COD and BOD) in the Kabul river in Afghanistan which severely affect aquatic life (Abdullah *et al.*, 2019). The rapid rise in population and urbanization and associated increased sewage discharge mean that the river can no longer offer the pollution dilution services it has provided up until now (Jawadi *et al.*, 2020).

The degradation of water quality in the Kabul river is not restricted to Afghanistan. Flow from the Swat river has been shown to contribute

Fig. 5.6. Kabul river canal, Peshawar district. (From: Amjad Jamal, IWMI-Pakistan)

to higher downstream concentrations of various contaminants and increased eutrophication (Barinova *et al.*, 2016). Not only does this lead to organic pollution increase but it also causes the trophic state downstream to become eutrophic and meso-eutrophic (Khuram *et al.*, 2019). Initial water quality modelling attempts have led to similar conclusions. Hydro-bacterial modelling results suggest that about 34% of the high levels of E. coli observed in water samples across cities in the Kabul basin in Pakistan originate from Afghanistan (Iqbal and Hofstra, 2018). This suggests that deterioration in streamflow quality in Afghanistan persists well downstream into Pakistan.

Water quality impacts are not restricted to surface water. In recent years, increasing levels of contaminants have emerged in basin groundwater. As expected, there is significant spatial variation in groundwater quality owing to variations in population, water use and urbanization (Mack *et al.*, 2013). Areas with lower population density have groundwater with low levels of total dissolved solids concentration. Conversely, high-density areas such as Kabul City show elevated levels of total dissolved solids, coliform bacteria and turbidity (Jawadi *et al.*, 2020). In areas with high irrigation intensity, high levels of nitrates and phosphates have been observed in groundwater (Saffi, 2011).

The existing literature thus paints a picture of a basin severely affected by human activities. The limited literature provides a reference point for location-specific impact but is unable to provide a holistic understanding of water quality on a basin level. Little is known about the health of the river beyond Kabul City, water use around it, and the spatio-temporal variations in water quality. As a result, the effectiveness of policies for improving water quality and ecosystem health is unclear.

Reservoirs and hydropower

There is relatively little water infrastructure in the Kabul river basin, including both storage reservoirs and hydropower facilities. Most water infrastructure is in the form of run-of-river hydropower facilities with limited storage. There are only two storage facilities

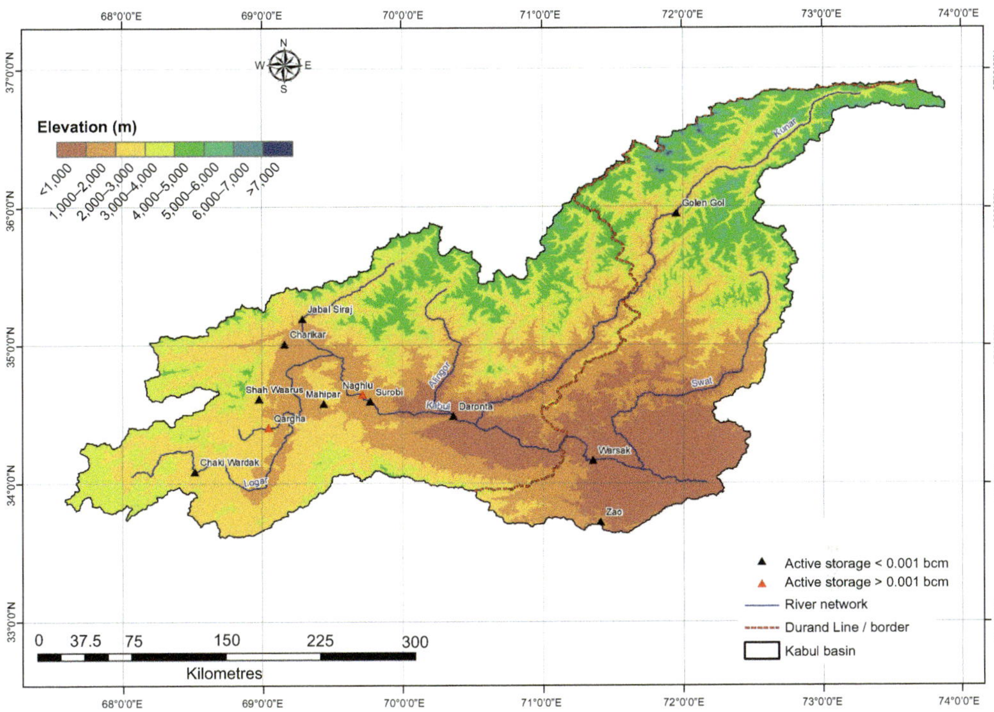

Fig. 5.7. Spatial distribution of existing water infrastructure in the Kabul river basin.

(Qargha and Naghlu) in the entire basin with an active capacity of over 1 mcm (both in Afghanistan), with cumulative storage in the basin comprising a small proportion of the average annual runoff (<5%). While there is little storage infrastructure in the Kabul river basin in Pakistan, there is significant irrigation infrastructure in KPK that makes use of Kabul river waters for the districts of Peshawar, Nowshera and Charsadda. Fig. 5.7. shows the spatial distribution of existing water storage infrastructure in the basin.

Owing to a lack of political stability and financial resources, limited water infrastructure development took place in Afghanistan during the latter part of the 20th century. However, since 2001, at least four storage reservoirs have been completed or are under construction in Afghanistan with the help of multilateral and bilateral donors. Most notable among these is the under-construction Shahtoot Dam, with a storage capacity of 147 mcm designed to meet municipal water demand in Kabul and agricultural demand in adjoining regions.[1] The projected increase in Afghanistan's Internally Displaced Persons returning from Pakistan and Iran and higher demands for food and energy may spur increased demand for water and necessitate additional water infrastructure (Najmuddin *et al.*, 2021), which could further impact flows from the Kabul river entering Pakistan.

There is significant untapped hydropower potential in both countries, especially in the transboundary Kunar sub-basin, something recognized in the ambitious future development plans of both Pakistan and Afghanistan's water resources institutions (Eqrar and Shroder, 2016; GoKP, 2016). In 2021, there were up to 36 potential water infrastructure projects identified in the basin across both Pakistan and Afghanistan. However, many of these projects remain at the conception stage and most have no feasibility assessment. It is unclear when or if any of these potential projects might be realized. It is also unclear how these infrastructure

Fig. 5.8. A stretch of the Kurram river, Mianwali district in Pakistan. (From: Amjad Jamal, IWMI)

developments (once completed), coupled with climate change, will affect food production and water quality in Afghanistan and downstream in Pakistan. The importance of the basin's surface flows extends well beyond the basin and deep into the Indus basin (downstream of Nowshera, Pakistan). Flows in the Kabul river peak earlier than in other tributaries of the Indus and thus are especially important for KPK and agriculture downstream in Sind that rely on these flows for the start of the summer-season cropping.

Recent research has explored the prospects of transboundary co-operation and conflict (Atef *et al.*, 2019) if the planned infrastructure development takes place. Jointly managed water infrastructure could lead to benefits for both countries, including increased winter energy generation (Khan *et al.*, 2020), augmented flows in the winter season, and greater resilience to hydrologic droughts (Taraky *et al.*, 2021). However, much more evidence is needed to adequately quantify the food, energy and water impacts across Afghanistan and Pakistan.

Kurram River Basin

Hydrology

The Kurram river originates in the province of Sayid Paktia and Khost in Afghanistan approximately 20 km south-east of Gardez (Fig. 5.9). From here, the river enters Pakistan at Kurram district in KPK. The river crosses the border south of the Kharlachi border post, south-west of Parachinar at a reduced distance (river mile) 158.[2] The Kurram river drains at Koh-e-Sufed mountains and joins the Indus river at Essa Khel in Mianwali district. The Kaitu river is a tributary of the Kurram river. The Kaitu river originates in Zadran province of Afghanistan, flows into Pakistan and joins the Kurram river approximately one mile north of Zarwam Village and downstream of the proposed Kurram-Tangi Dam. The catchment of the Kurram river above the location of the proposed Kurram-Tangi Dam is approximately 11,300 km^2. The Kurram is usually narrow and shallow

Fig. 5.9. Kurram river basin.

Table 5.2. Streamflow-gauging stations in the Kurram river basin.

Station	Location	River	Data start	Data end
Thal	Mile 96 (Thal)	Kurram river	1968	To date
Spinwam	Mir Ali to Thal road bridge	Kaitu river	1979	To date
Kurram Garhi Headworks	Mile 64	Kurram river	1958	To date

but prone to torrential flows from its headwaters in Afghanistan. There are three flow-gauging stations in the Kurram river basin (Table 5.2).

Average flow volumes (1980–2011) at gauging stations in the Kurram river basin show notable variation (Fig. 5.10). The hydrographs are not dissimilar to those of the Kabul river basin and typical for rivers of the upper Indus basin. Streamflow in the basin is strongly seasonal, with pronounced snow and glacier melt in the spring and summer seasons as a contributing source.

Seasonal flow variation

Streamflow peaks in all the locations in the spring and summer months between March and September and the volumes are orders of magnitude smaller than those of the Kabul river basin. The highest monthly volumes recorded at the Thal, Kurram Garhi Headworks and Spinwam peak in the range of 120–180 mcm, whereas the Kabul river basin at Nowshera peaks at nearly 5 bcm in July.

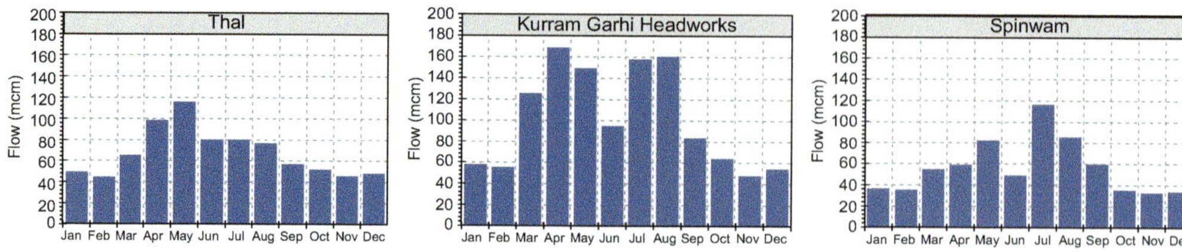

Fig. 5.10. Average flow volumes in million cubic metres, 1980–2011, Kurram river basin. (Adapted from WAPDA, 2011)

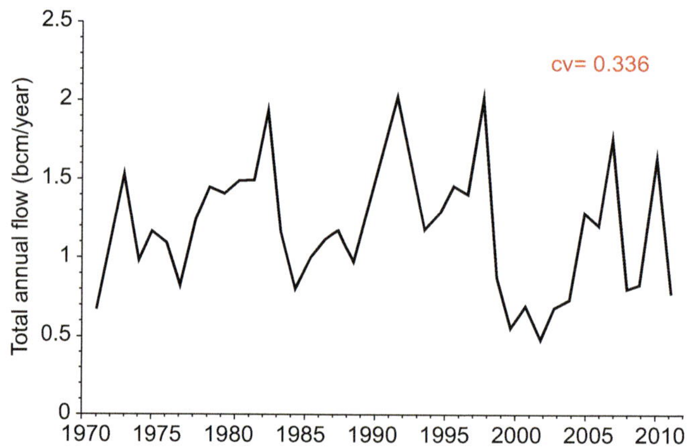

Fig. 5.11. Total annual streamflow measured at Kurram Garhi, the most downstream gauge in the Kurram river basin.

Inter-annual flow variation

Fig. 5.11 shows the time series of annual discharge for Kurram Garhi Headworks streamgauge station for records from 1971–2011. The records show that the average annual flow is just under 1.19 bcm/year over 40 years of recorded flows at the Kurram Garhi Headworks gauge station. In 40 years, the minimum of 0.47 bcm/year was observed with a maximum value of 2 bcm/year. Fluctuations in annual flow were observed between 0.5 bcm/year and 1.5 bcm/year.

Impact of climate change

The impact of climate change on the Kurram river basin has received little attention. WAPDA (2011) does not consider the impact of climate change. MWH Americas, Inc. (MWH) was the only significant attempt to investigate the impact of climate change. MWH, 2013b used ordinary least squares regression on flow records from the Thal station (presumably observed rather than synthesized but unspecified) and concludes that the annual volumes at this station are decreasing by 5 mcm annually. However, flows at Kurram-Tangi gauge station are increasing by 2 mcm and at Spinwam by 7 mcm annually. The inconsistency of these results illustrates the challenge of applying a simple ordinary least squares regression to a data set and ultimately calls for caution when characterizing flow trends in the Kurram river basin. There was no attempt by MWH to determine if time was a significant explanatory variable of flow.

MWH, 2013c also applied 18 global climate change models (GCMs) to assess possible

Table 5.3. Surface water quality in the Kurram river basin. (Adapted from MWH, 2013a)

Location	Electrical conductivity (dS/m)	Sodium absorption ratio	Residual sodium carbonate (meq/l)	pH	Chloride (meq/l)
Kaitu river, Spinwam Bridge	890	1.60		7.90	0.70
Kaitu river, Spinwam Bridge (May 2003)	1000	1.80	0	7.80	1.10
Kurram River Dam site	460	0.80	0	7.80	0.50
Kurram Garhi Headworks	570	0.50		8.00	0.70
Tochi river	770	1.10		8.20	1.60
Baran Dam	550	0.90		8.00	0.80
Kashu Algard	8740	34.0		8.00	72
Marwat canal at outlet	510	1.3		8.00	0.70
Marwat canal at farm gate	520	1.40		8.00	0.80
Bannu-Kohat road crossing	460	0.40	0	7.80	0.60
Domel-Azim Kele road crossing	1060	36.2	0	7.80	81.00

future changes in precipitation and temperature in this region. Due to the poor correlation, they did not use absolute values but rather looked at the change predicted by the various GCMs and reported that the ensemble of these models predicted little change in precipitation (4% by 2099, which is well within a decadal range of variation). MWH used the same three GCMs to predict temperature changes in the future and here the models provided much higher correlations. An ensemble of these models suggested warming for Parachinar, primarily in the summer months, and a possible increase of 3.4°C by 2065 and 5.2°C by 2099. Similar temperature increases were reported for the town of Bannu in north-west Pakistan.

Water quality

Water quality monitoring in the Kurram river basin is limited and comprises mostly spot measurements around the proposed Kurram-Tangi Dam for feasibility studies and allied irrigation systems. Table 5.3 summarizes chemical water quality parameters for spot measurements in the river basin. Overall, water quality is in need of improvement and the electrical conductivity values are on the higher side.

Sediment load in the Kurram river is also a matter of concern. WAPDA (2011) cites the sediment load at Thal gauge station as 705 m^3 per km^2 of drainage area measured by the Surface Water Hydrology Project but offers little in the form of critique or cross-checking of this estimate. MWH, 2013b noted the contradiction that although the flow measurements at Thal gauge station have been questioned, the sediment measurements at Thal stream station have been accepted without question. They further note that Surface Water Hydrology (SWH)'s publication *Sediment Appraisal of Pakistan Rivers 1960–2005* provides an annual sediment yield of Kurram river at Thal as 742 m^3 per km^2, which is slightly higher than that in WAPDA (2011) of 705 m^3 per km^2. Furthermore, they plot sediment yield (normalized by catchment area) against annual runoff (normalized by catchment area) for seven locations with comparable river basins. Using a log-linear regression, they estimate that the sediment yield at Thal station is 1376 m^3 per km^2. This estimate is highly questionable since the sediment yield of a catchment is a complex phenomenon and cannot be explained by annual runoff alone. The p-value of annual runoff is very high (p = 0.316) suggesting this explanatory variable has little explanatory power for sediment yield.

Reservoirs and hydropower

The Kurram river basin remains largely undeveloped. The Kurram Garhi Headworks and the downstream irrigation systems (Marwat canal, Bannu civil canals) are the only notable water resources infrastructure in this river basin and date from 1962. These canals originally irrigated 112,550 ha, although this capacity has reduced somewhat. The Baran river is a tributary of the Kurram river. The Baran Dam, completed in 1963, impounds the water of the Baran river and is the only existing dam in this basin. By 2020, the dam had largely silted up and further investment is being made to raise it to extend its service life (Saeed and Khan, 2014).

The Kurram-Tangi Integrated Water Resources Development Project has gone through numerous iterations. It is reported to have been conceptualized as early as 1955 (WAPDA, 2004) envisaging the development of natural and human resources (land, water, labour) for social, economic and environmental outcomes. The proposed Kurram-Tangi Dam will impound the Kurram river and create a reservoir with a storage capacity of 1.48 bcm. Whereas the proposed dam is at a reduced distance (river mile) of 96. Fig. 5.12 is a schematic of the Kurram-Tangi Integrated Water Resources Development Project. A weir (Kaitu weir) is under construction along the Kaitu river, and a feeder tunnel and open channel between the Kaitu river at this weir location will allow for managed flows between the Kaitu river and the Kurram reservoir. An 18 MW power station shown as PH-IV in Fig. 5.12 is proposed on the feeder between Kaitu weir and Kurram reservoir, hence the feeder tunnel/canal is often called the Kaitu power feeder. There are two new irrigation canals proposed from the Kaitu weir, the Sheratalla canal irrigating 5000 ha and Spaira Ragha canal irrigating another 5700 ha.

Gomal River Basin

Hydrology

The Gomal river basin is the smallest of the three with a catchment area of 39,200 km^2. The Gomal river contributes a relatively modest 970 mcm/year (Bhatti *et al.*, 2021). The Gomal river catchment upstream of the Kajuri Kach gauge station is 20,200 km^2 (Fig. 5.13). The catchment upstream of Kot Murtaza barrage from where the Gomal Zam irrigation system offtakes are located is roughly equivalent to the entire size of the basin (Bhatti *et al.*, 2021). The two provinces of Baluchistan and KPK in Pakistan contribute to 70% of the Gomal Zam catchment. The area is arid with a reported rainfall of only 230 mm/year, whereas total evaporation is approximately 1900 mm/year. Net evaporation of rainfall is in the order of 1600 mm. There are two gauging stations in the Gomal river basin, both within Pakistan (Fig. 5.13).

Average flow past various gauge stations in the Gomal Zam river basin illustrate two peaks (Fig. 5.14). The hydrographs exhibit two pronounced spikes in spring and summer. The time series for Kajuri Kach and Kot Murtaza are nonetheless not identical, which may explain why the peak flow months are slightly different. The peak volume at both gauging stations is broadly comparable. This is expected as there is only a 35 km reach of the Gomal Zam river between the two gauging stations with relatively few substantial tributaries. In off-peak months, the volume at Kot Murtaza is greater than that at Kajuri Kach. A precise explanation for this cannot be confirmed; it may be that the smaller tributaries contribute more in the off-peak months to the volumes at Kot Murtaza.

WAPDA (2014) reports that the annual average water available at Kajuri Kach is 1030 mcm/year based on flow records from 1962–1995, and for Kot Murtaza 770 mcm/year based on flow records from 1960–1990 (Fig. 5.15; Table 5.4). This is counter-intuitive as one would expect higher volumes at the downstream gauge station of Kot Murtaza. The water resource as documented in WAPDA (2014) is broadly comparable with the figure reported by Bhatti *et al.* (2021) of 974 mcm/year.

Water quality

Water quality testing in the Gomal river basin is limited. Muhammad and Ullah (2022) reported on the surface water quality at various locations around the reservoir upstream of the Gomal Zam

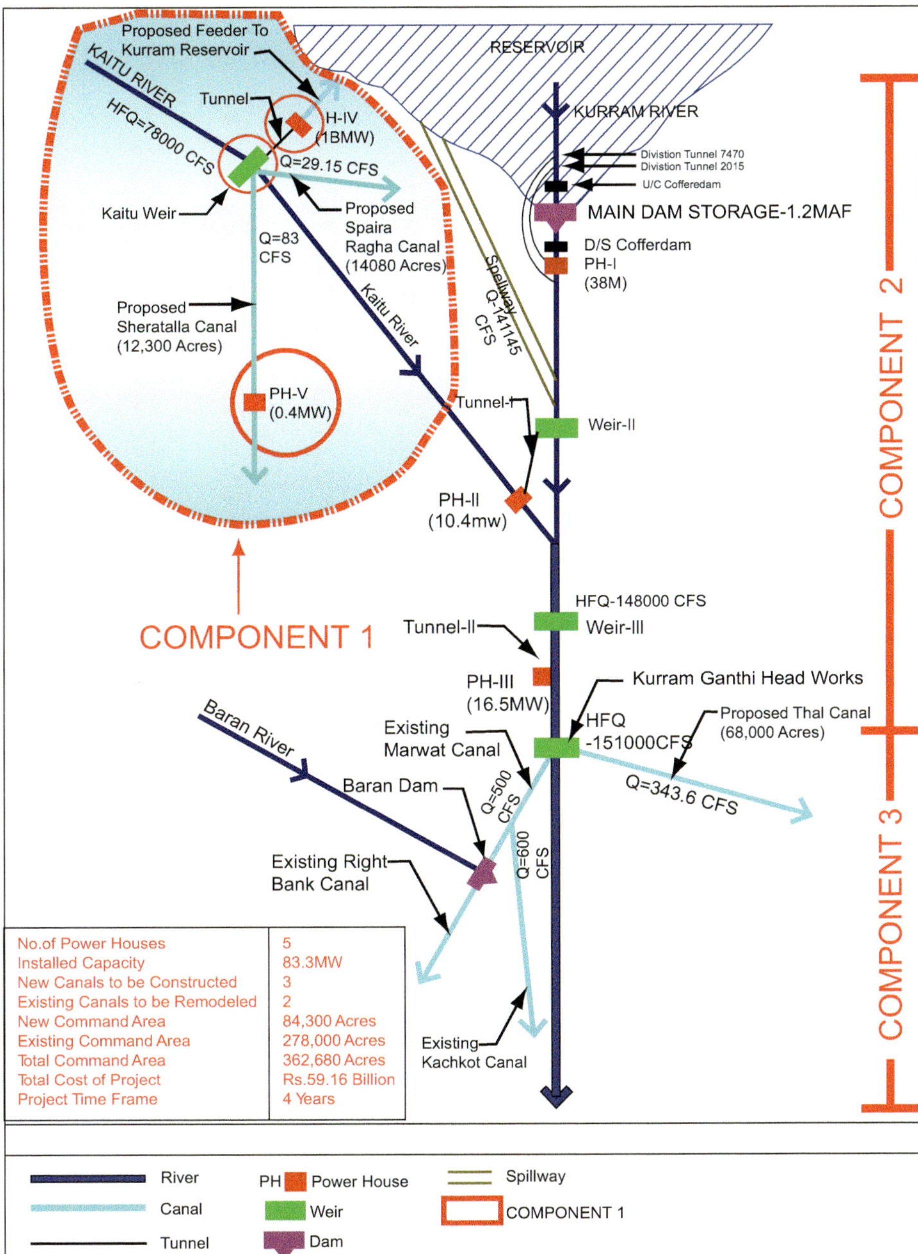

Fig. 5.12. Kurram-Tangi Integrated Water Resources Development Project. (Adapted from MWH, 2013b)

Dam in both the winter and summer seasons. They concluded that the water was generally of good quality within WHO guidelines other than for turbidity. The Water Quality Index parameters are in the 'good' category for drinking purposes. Similarly, for irrigation, the Gomal Zam Dam and its tributary waters were classified in the 'permissible' range. WAPDA (2014) also reported on the Gomal Zam water quality for construction purposes. Turbidity was reported as 78 NTU and specific conductivity 906 μS/cm. Chemical characteristics are within permissible limits. Water quality testing in the

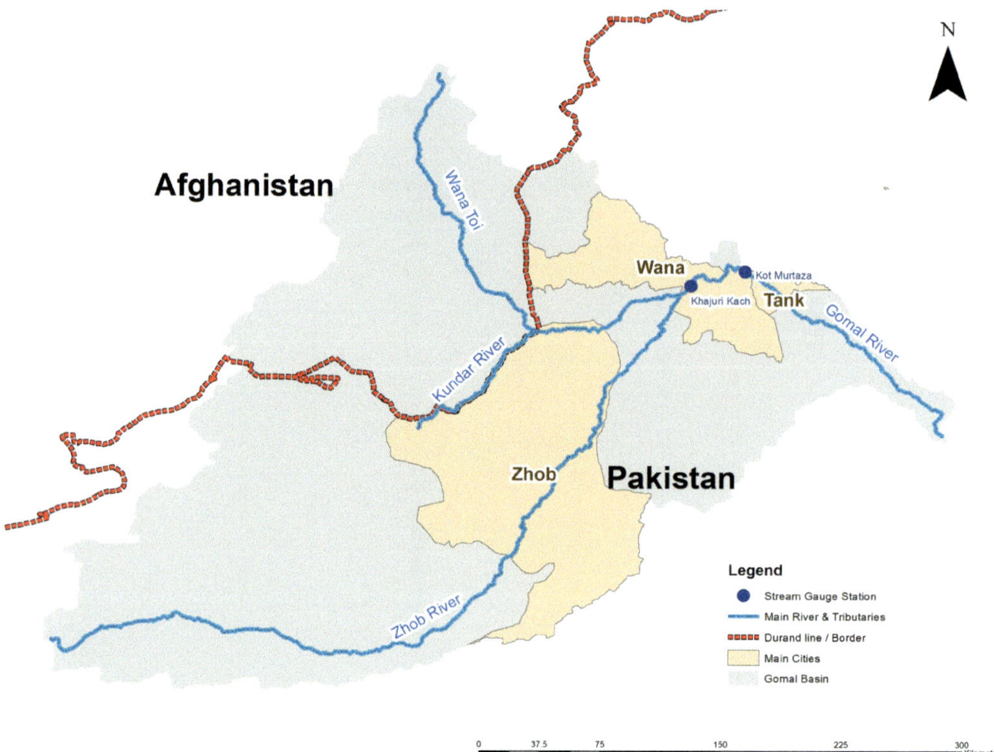

Fig. 5.13. Gomal river basin.

Table 5.4. Streamflow gauging stations in the Gomal river basin.

Station	River	Location	Data years
Kajuri Kach	Gomal river	Lat. 32°4'0" Long. 69°51'8"	Kajuri Kachi (1994, 1995, 1996 incomplete, 2009 incomplete, 2011)
Kot Murtaza	Gomal river	Lat. 32°9'15" Long. 70°4'9"	Kot Murtaza (2002 incomplete, 2003–2005, 2006 incomplete, 2006–2009)

Gomal Zam basin is very limited. However, with the data available, overall, the ground and surface waters in the Gomal Zam basin are well within permissible limits with turbidity the only major concern.

Reservoirs and hydropower

The only significant infrastructure is the Gomal Zam Dam and its allied irrigation system. There are no plans for any further investment in the Gomal river basin. The Gomal Zam Dam is in South Waziristan district of the north-west province of KPK in Pakistan. The nearest town, Wana, is approximately 35 km from the dam. The nearest city, Dera Ismail Khan, is about 150 km away. The dam impounds the Gomal river and Zhob river (a tributary of the Gomal) at Khajuri Kach. The dam is a roller-compacted concrete dam approximately 140 m high and serves multiple purposes (irrigation, flood control and power generation) (Wang *et al.*, 2014). The reservoir behind the dam has a total capacity of 2 bcm and the dam has an installed hydroelectric capacity of 17.4 MW (WAPDA,

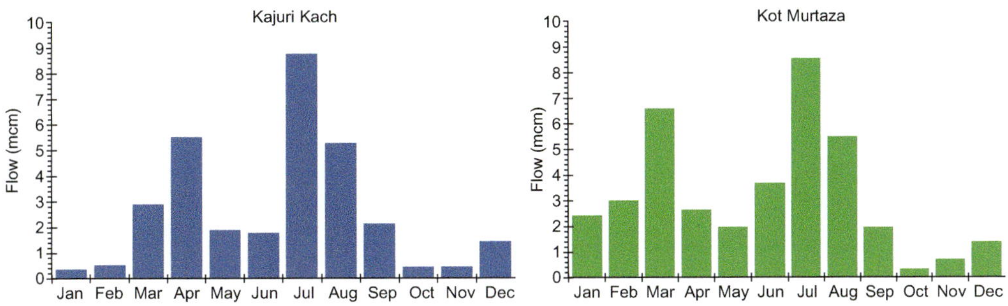

Fig. 5.14. Volume flowing past gauge stations at (a) Kajuri Kach and (b) Kot Murtaza. (Adapted from WAPDA, 2011)

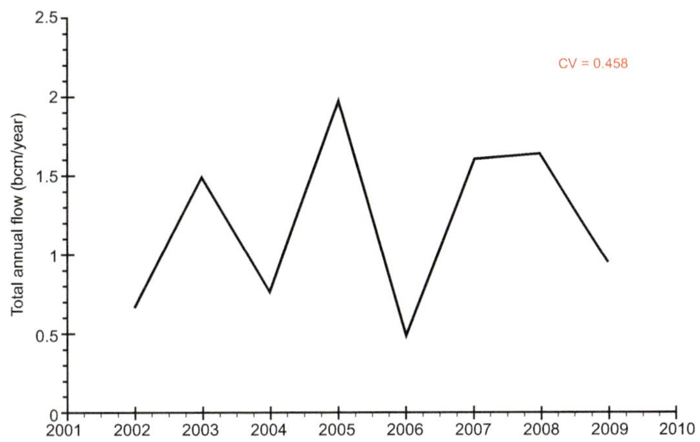

Fig. 5.15. Total annual streamflow measured at Kot Murtaza, the most downstream gauge in the Gomal river basin.

2014). The irrigation system downstream of the dam is designed to irrigate approximately 70,000 ha in Tank, Kulachi and Dera Ismail Khan districts of KPK (Khan *et al.*, 2018).

Conclusions

Above and beyond the specifics of inter- and intra-annual flow variability, which appear to be increasing, the data suggest at least four main messages. First, water-storage infrastructure is currently low in these basins. Second, irrigation-water infrastructure is not low, particularly in Pakistani portions of the Kabul basin, which engenders a reasonable level of dependence on the rivers. Third, water quality, particularly in the Kabul river, is a cause for concern and could be putting populations at risk. Fourth, monitoring is sparse and data gaps on basic parameters like seasonal flow and sediment yield loom large.

As noted in Chapter 3, this region remains desperately poor and needs investment for socio-economic development. Water infrastructure development is one clear vehicle for socio-economic improvement and, accordingly, both countries have made efforts to expand water infrastructure. Pakistan has independently developed its Indus basin irrigation system and has recently developed water infrastructure on the Gomal river and infrastructure development is underway on the Kurram river. Afghanistan has similar ambitions, particularly for the Kabul and Kunar rivers, although the instability that has followed the 2021 government change may slow progress. Such efforts are nonetheless being made suboptimally for two reasons: (i) water planning and investment is made unilaterally; and (ii) these investments are being made in the context of inadequate data and information. In light of these limitations on current development efforts, this chapter proposes five key recommendations to guide an improved pathway forward:

- **Co-ordinate across countries in water resources assessment and planning, at a minimum**. While achieving active and ongoing joint management of shared water resources is a noble goal, such an aim may not be achievable in the near term in the Kabul, Kurram and Gomal river basins. Nonetheless, even if management is separately implemented by the two countries, there would appear to be key opportunities for collaborative water assessments through, for example, data exchange to strengthen management within the countries. Improved knowledge based on the hydrology of the full basin can inform planning and strengthen management at a national level. The integration of hydrological data in this chapter indeed provides an example of the fuller picture that can be generated through such an effort.
- **Inform development efforts through data collection, synthesis and analysis**. As noted above, the knowledge base on which water infrastructure is designed and constructed, and indeed on which water development decisions are made, is inadequate. There is a need for more data that can be processed to support development decisions and infrastructures. Any such effort requires focus to ensure that data collected are plugged into a process that supports decisions. Achieving development in an optimal way requires data collection in an optimal way.
- **Develop sediment and soil erosion models**. There are several tools that can estimate soil erosion. The universal soil loss equation (USLE) is one of the most widely used. Any of these soil erosion models, whether physical or empirical, are not without their shortcomings. The best option might be to apply an ensemble of models to estimate the sediment yield which would be a more accurate method of assessment.
- **Monitoring efforts need to be sustained and expanded**. The Kabul is the only river where we have better monitoring of surface water although data gaps are present due to episodes of war and instability on the Afghan side. The Kurram and Gomal rivers do not have adequate monitoring infrastructure on both sides of the Durand Line. This presents another opportunity for both countries to develop a proper monitoring infrastructure to safeguard their current and planned infrastructure developments in these basins.
- **Explore private sector participation in monitoring**. Clearly, it is important to institute a mechanism for monitoring of surface and groundwater in Afghanistan and Pakistan. The catchments of the Kurram and Gomal rivers are in remote areas along the border and physical monitoring of gauges seems to be a difficult task. One way to address this issue is to involve the private sector to install automatic monitoring stations that require limited human interventions. Involvement of the private sector may support maintenance of the installed infrastructure as payments for actors in the sector could be linked to the provision of continuous availability of data.

Notes

1 See for example: https://thediplomat.com/2018/08/indias-controversial-afghanistan-dams/; https://www.livemint.com/news/india/india-and-afghanistan-may-sign-pact-on-300-mn-dam-this-week-11612805733668.html; https://foreignpolicy.com/2018/11/13/afghanistans-rivers-could-be-indias-next-weapon-against-pakistan-water-wars-hydropower-hydrodiplomacy/

2 Reduced distances in rivers (and natural drains) are measured linearly from the outfall upstream, whereas reduced distances in canals are measured from the head downstream.

References

Aawar, T. and Khare, D. (2020) Assessment of climate change impacts on streamflow through hydrological model using SWAT model: a case study of Afghanistan. *Modeling Earth Systems and Environment* 6(3), 1427–1437. DOI: 10.1007/s40808-020-00759-0.

Abdullah, X., Zhou, D., Shah, T., Ali, S., Ahmad, W. *et al.* (2019) Factors affecting household food security in rural northern hinterland of Pakistan. *Journal of the Saudi Society of Agricultural Sciences* 18(2), 201–210. DOI: 10.1016/j.jssas.2017.05.003.

Akhtar, S.M. and Iqbal, J. (2017) Assessment of emerging hydrological, water quality issues and policy discussion on water sharing of transboundary Kabul River. *Water Policy* 19(4), 650–672. DOI: 10.2166/wp.2017.119.

Atef, S.S., Sadeqinazhad, F., Farjaad, F. and Amatya, D.M. (2019) Water conflict management and cooperation between Afghanistan and Pakistan. *Journal of Hydrology* 570, 875–892. DOI: 10.1016/j.jhydrol.2018.12.075.

Baig, M.H.A., Abid, M., Khan, M.R., Jiao, W., Amin, M. *et al.* (2020) Assessing meteorological and agricultural drought in Chitral Kabul River Basin using multiple drought indices. *Remote Sensing* 12(9), 1417. DOI: 10.3390/rs12091417.

Barinova, S., Khuram, I., Ahmad, N., Jan, S, *et al.* (2016) How water quality in the Kabul River, Pakistan, can be determined with algal bio-indication. *Advanced Studies in Biology* 8, 151–171. DOI: 10.12988/asb.2016.6830.

Bhatti, M.T., Ashraf, M. and Anwar, A.A. (2021) Soil erosion and sediment load management strategies for sustainable irrigation in arid regions. *Sustainability* 13(6), 3547. DOI: 10.3390/su13063547.

Bokhari, S.A.A., Ahmad, B., Ali, J., Ahmad, S., Mushtaq, H. *et al.* (2018) Future climate change projections of the Kabul River Basin using a multi-model ensemble of high-resolution statistically downscaled data. *Earth Systems and Environment* 2(3), 477–497. DOI: 10.1007/s41748-018-0061-y.

Eqrar, N. and Shroder, J. (2016) Development of water resources in the Kabul River Basin. *Transboundary Water Resources in Afghanistan*, 91–119. DOI: 10.1016/b978-0-12-801886-6.00004-5..

GoKP (Government of Khyber Pakhtunkhwa) (2016) Khyber Pakhtunkhwa hydropower policy. Available at: https://pedo.pk/uploads/policies/KP_Hydropower_Policy_2016.pdf (accessed 24 July 2022).

Hashmi, M.Z., Masood, A., Mushtaq, H., Bukhari, S.A.A., Ahmad, B, *et al.* (2020) Exploring climate change impacts during first half of the 21st century on flow regime of the transboundary kabul river in the hindukush region. *Journal of Water and Climate Change* 11(4), 1521–1538. DOI: 10.2166/wcc.2019.094.

Iqbal, M.S. and Hofstra, N. (2018) Modeling Escherichia coli fate and transport in the Kabul River Basin using SWAT. *Human and Ecological Risk Assessment: An International Journal* 25(5), 1279–1297. DOI: 10.1080/10807039.2018.1487276.

Iqbal, M.S., Dahri, Z., Querner, E., Khan, A. and Hofstra, N. (2018) Impact of climate change on flood frequency and intensity climate change on flood frequency and intensity in the Kabul River Basin. *Geosciences* 8(4), 114. DOI: 10.3390/geosciences8040114.

Iqbal, M.S., Islam, M.M.M. and Hofstra, N. (2019) The impact of socio-economic development and climate change on E. coli loads and concentrations in Kabul River, Pakistan. *Science of the Total Environment* 650(Pt 2), 1935–1943. DOI: 10.1016/j.scitotenv.2018.09.347.

Jawadi, H.A., Sagin, J. and Snow, D.D. (2020) A detailed assessment of groundwater quality in the Kabul Basin, Afghanistan, and suitability for future development. *Water* 12(10), 2890. DOI: 10.3390/w12102890.

Khan, S.A., Shah, S.A., Ullah, I., Ibrahim, M. and Khan, S. (2018) Impact of Gomal Zam Dam irrigation project on agriculture and welfare of farming community in southern districts of Khyber Pakhtunkhwa-Pakistan. *Asian Journal of Agriculture and Rural Development* 7(10), 212–218. DOI: 10.18488/journal.1005/2017.7.10/1005.10.212.218.

Khan, H.F., Yang, Y.C.E. and Wi, S. (2020) Case study on hydropolitics in Afghanistan and Pakistan: energy and water impacts of Kunar River development. *Journal of Water Resources Planning and Management* 146(9), 05020015. DOI: 10.1061/(ASCE)WR.1943-5452.0001262.

Khattak, M.S., Remant, N.U., Sharif, M. and Khan, M.A. (2015) Analysis of streamflow data for trend detection on major rivers of the Indus basin. *Journal of Himalayan Earth Sciences* 48(1), 99–111.

Khuram, I., Muhammad, Z., Ahmad, N., Ullah, R. and Barinova, S. (2019) Green and charophyte algae in bioindication of water quality of the Shah Alam River (District Peshawar, Pakistan). *Transylvanian Review of Systematical and Ecological Research* 21(1), 1–16. DOI: 10.2478/trser-2019-0001.

Latif, Y., Ma, Y. and Ma, W. (2021) Climatic trends variability and concerning flow regime of Upper Indus Basin, Jehlum, and Kabul river basins Pakistan. *Theoretical and Applied Climatology* 144(1–2), 447–468. DOI: 10.1007/s00704-021-03529-9.

Mack, T.J., Chornack, M.P. and Taher, M.R. (2013) Groundwater-level trends and implications for sustainable water use in the Kabul Basin, Afghanistan. *Environment Systems and Decisions* 33(3), 457–467. DOI: 10.1007/s10669-013-9455-4.

Mehmood, A., Jia, S., Mahmood, R., Yan, J. and Ahsan, M. (2019) Non-stationary Bayesian modeling of annual maximum floods in a changing environmentand implications for flood management in the Kabul River Basin, Pakistan. *Water* 11(6), 1246. DOI: 10.3390/w11061246.

Mehmood, A., Jia, S., Lv, A., Zhu, W., Mahmood, R, *et al*. (2021) Detection of spatial shift in flood regime of the Kabul River Basin in Pakistan: causes, challenges, and opportunities. *Water* 13(9), 1276. DOI: 10.3390/w13091276.

Muhammad, S. and Ullah, I. (2022) Spatial and seasonal variation of water quality indices in Gomal Zam Dam and its tributaries of south Waziristan District, Pakistan. *Environmental Science and Pollution Research International* 29(19), 29141–29151. DOI: 10.1007/s11356-022-18524-4.

MWH (2013a) *Kurram Tangi Dam Construction, Environmental Assessment*. Vol. I: Main Report (Final). MWH Americas Inc., Bloomfield, Colorado.

MWH (2013b) *Kurram Tangi Dam Construction Supplemental Report on Hydrology*. Final Report. MWH Americas Inc, Bloomfield, Colorado. Available at: https://pdf.usaid.gov/pdf_docs/PA00KB13.pdf (accessed 24 July 2022).

MWH (2013c) *Kurram Tangi Dam Construction Supplemental Report on Climate Change*. Final Report. MWH Americas Inc, Bloomfield, Colorado.

Najmuddin, O., Qamer, F.M., Gul, H., Zhuang, W. and Zhang, F. (2021) Cropland use preferences under land, water and labour constraints: implications for wheat self-sufficiency in the Kabul River basin, Afghanistan. *Food Security* 13(4), 955–979. DOI: 10.1007/s12571-021-01147-x.

Saffi, H. (2011) *Groundwater Natural Resources and Quality Concern in Kabul Basin*. International Symposium for Managed Aquifer Recharge, Beijing, China.

Saeed, T.U. and Khan, D. (2014) Assessment and conservation of groundwater quality: a challenge for agriculture. *British Journal of Applied Science & Technology* 4(8), 1256.

Taraky, Y.M., McBean, E., Liu, Y., Daggupati, P., Shrestha, N.K. *et al*. (2021) The role of large dams in a transboundary drought management co-operation framework: case study of the Kabul River Basin. *Water* 13(19), 2628. DOI: 10.3390/w13192628.

Taraky, Y.M., Liu, Y., Gharabaghi, B., McBean, E., Daggupati, P. *et al*. (2022) Influence of headwater reservoirs on climate change impacts and flood frequency in the Kabul River Basin. *Canadian Journal of Civil Engineering*.

Wang, G., Hu, W. and Duffield, C. (2014) Risk mitigation in managing a mega project-a case study of Gomal Zam Dam construction. *Journal of Risk Analysis and Crisis Response* 4(1).

WAPDA (2004) *Kurram Tangi Dam Project*. Feasibility Report, Main Report. Water and Power Development Authority, Peshawar, Pakistan.

WAPDA (2011) *Detailed Design Report*. Chapter 2: Hydrology and sedimentation. Water and Power Development Authority, Peshawar, Pakistan.

WAPDA (2014) *Gomal Zam Multipurpose Project*. PC-1. Water and Power Development Authority, Peshawar, Pakistan.

Wi, S., Yang, Y.C.E., Steinschneider, S., Khalil, A. and Brown, C.M. (2015) Calibration approaches for distributed hydrologic models in poorly gaged basins: implication for streamflow projections under climate change. *Hydrology and Earth System Sciences* 19(2), 857–876. DOI: 10.5194/hess-19-857-2015.

6 Groundwater

Muhammad Shahid Iqbal and Paul Pavelic

Chapter Overview

This chapter presents an overview of the groundwater resources of the upper Indus basin's three major transboundary river basins, with particular attention to the Kabul river basin, where information is more readily available. It examines the state of knowledge and information, challenges, and gaps in and barriers to sustainable groundwater resource management. Afghanistan and Pakistan are highly water-scarce countries that have traditionally not given adequate priority to managing their groundwater resources. This has resulted in a situation of inadequate scientific knowledge, technical capacity, policies and regulatory frameworks in relation to groundwater. Recommendations to address these gaps are provided.

Fig. 6.1. Groundwater pumping-solar irrigation, command area of Gomal Zam Dam. (From: Kashif Hussain, IWMI-Pakistan)

DOI: 10.1079/9781800622371.0006

Introduction

Groundwater resources play a key role in the South Asia region, where the resource is of critical importance for the day-to-day needs of around 1.5 billion people (Hirji *et al.*, 2017; World Bank, 2020).[1] Reliance on groundwater is expected to increase even further. The high dependency on groundwater is strongly related to the region's climate, characterized by a high degree of uncertainty in the timing and magnitude of seasonal rainfall and a protracted dry season. Groundwater will play an even more prominent role in water resources planning and management to advance climate resilience and satisfy growing water demand among populations, economies, industries and cities (Satoh *et al.*, 2017). If managed well, groundwater resources can provide a powerful means of climate change adaptation and contribute positively towards many sustainable development goals (SDGs) (GGP, 2016).

In Afghanistan and Pakistan, specifically, water resource scarcity is anticipated to become increasingly problematic as precipitation patterns change across the region. Due to climate change, short- and long-term droughts are becoming more frequent and intense (CDKN, 2014). The threat of drought to irrigated farming is already very clear. In 2014, for example, an estimated 31% of the land that is normally available for irrigation across Pakistan was not used because of below-average precipitation (–12.6%) and hence there was a lack of sufficient water in the rivers and irrigation canal systems (NDMC, 2014). Irrigated areas in Afghanistan have been hit hard by reduced snowfall in the Hindu-Kush mountain range in spring.[2] Coupled with rapid population increases, climate change will place pressure on regional water resources, particularly groundwater, as precipitation becomes more erratic and rising temperatures increase water demand. Ultimately, this has led to estimates that 50% of water demand may go unsatisfied by 2100 (World Bank, 2021).

Effective management and protection of groundwater, of critical importance to alleviating present and looming scarcity and climate change challenges, must be informed by scientific data and analysis. In basins shared by Afghanistan and Pakistan, however, knowledge of groundwater is not extensive and is universally approached from the perspective of one of the two riparian countries. For example, Qureshi (2011) explores the unsustainable use of groundwater for agriculture in Pakistan. Due to constraints in surface water availability, farmers started switching to groundwater, causing groundwater abstraction from the private wells installed in Punjab to increase to 51 x 10^9 m^3 as compared to recharge of 40–60 x 10^9 m^3. Uhl (2003) describes the hydrogeological systems and groundwater use in Afghanistan. Whilst an integrated assessment of groundwater resources has not been conducted across the three basins, there has been some examination of groundwater abstraction at the scale of the broader South Asia region (e.g. Cheema *et al.*, 2014).

The authors of this chapter compiled and synthesized current knowledge on the groundwater resources for three major transboundary sub-basins in the upper Indus river basin shared between Afghanistan and Pakistan; the Kabul, Kurram and Gomal. While we did not undertake a new analysis, we believe this to be the first effort to integrate groundwater data and information from the two countries on these three basins and to support considerations of how groundwater can be included in water management at a basin level in the Kabul, Kurram and Gomal. The chapter first describes the level of groundwater data and information available in the three shared basins. The next sections address each of the three core elements of groundwater: (i) hydrogeology; (ii) recharge; and (iii) abstraction. Finally, the chapter considers groundwater management holistically and provides recommendations. Given the disproportionate data availability, greater focus is devoted to groundwater in the Kabul basin.

Groundwater Data Existence and Compilation

The region is characterized by an overall scarcity of data, which is particularly acute in the Kurram and Gomal basins where little or no information exists. Hydrogeological studies conducted in Afghanistan during the 2000s and 2010s by international agencies including the United States Geological Survey (USGS) and the Federal

Institute for Geosciences and Natural Resources (BGR) have largely focused on the Afghanistan part of the Kabul river basin (e.g. Broshears *et al.*, 2005; Houben *et al.*, 2009; Mack *et al.*, 2013). In more recent years, national agencies have built upon these earlier works (e.g. Saffi, 2019).

In Afghanistan, monitoring networks are in place and maintained by the Ministry of Energy and Water, yet appear to be limited to the Kabul region. Groundwater use is not metered and estimates of abstractions reported here are based on crude estimates. There appears to be limited data in sedimentary and hard rock hydrogeological environments. In Pakistan, the provinces of Punjab and Sindh have installed monitoring networks in collaboration with the Pakistan Council for Research in Water Resources (PCRWR) and these are maintained by the respective provincial governments. The Department of Irrigation and Public Health is responsible for monitoring and evaluating groundwater-level data. In Khyber Pakhtunkhwa province (KPK), there are as yet no monitoring wells to estimate groundwater storage. The United States Agency for International Development (USAID) in collaboration with PCRWR is planning to install monitoring wells across the entire province, especially covering the Kabul, Kurram and Gomal basins, to help monitor groundwater levels and quality at the same time.

In this chapter, datasets provided in publications were heavily relied upon. The reliability of this data is presumed to be adequate given that collection was by trained groundwater specialists and the data was published by highly respected institutions and in peer-reviewed journals. Where necessary, data has been digitized from those reports. Raw data were unfortunately scarce and difficult to access.

Hydrogeology

The aquifers of the Kabul, Kurram and Gomal basins may be defined according to the geological units as identified by Wandrey and Law (1998) (Fig. 6.2). Productive aquifers are associated with the unconsolidated sediments primarily of the Quaternary age as well as the consolidated sediments of Tertiary and older periods. In the Kabul basin, unconsolidated aquifers extend across 11% of the basin and sedimentary rocks, 3% mainly in the southern parts. For the Kurram, the corresponding figures are 30% and 13%; for the Gomal the figures are 19% and 11%. The hard-rock aquifer systems are barely explored, and whilst they may offer important sources for domestic and irrigation supply in the future, for this chapter they are given limited attention due to the limited information available. As indicated in Fig. 6.2, unconsolidated sediments extend across the Durand Line in all three basins, indicating the potential for four transboundary aquifers. A clearer delineation of these aquifers would require more detailed, collaborative investigations between the two countries.

The Kabul river basin in Afghanistan has several key unconsolidated aquifers (Fig. 6.2). All these aquifers are generally comprised of sandy-gravel deposits formed by river terraces. These deposits are often coarse-grained, poorly sorted and frequently exhibit considerable variability. The gravel bodies work as a single hydraulic system without larger internal dividing or confining layers. Secondary mineralization processes frequently partially cement the pore spaces of deeper aquifers (Houben *et al.*, 2009).

The Kabul basin in Afghanistan can be characterized by valleys that are in-filled with Quaternary and Tertiary (Neogene) terrestrial and lacustrine sediments, primarily consisting of marls, and inter-bedded sand and gravel deposits that are bound by mountain ranges composed of uplifted crystalline, igneous and sedimentary rocks, as illustrated in Fig. 6.3 (Uhl, 2003; Broshears *et al.*, 2005; Mack, 2013).

Major sedimentary aquifers within the basin consist of coarse sandy to gravelly detritus deposited by major rivers flowing through the basin and originating from the surrounding mountains. These deposits are typically coarse-grained, poorly sorted and heterogeneous due to limited distances of transportation (Houben *et al.*, 2009). Quaternary sediments are typically less than 100 m thick, whereas the underlying Tertiary sediments may be up to 1000 m thick in places.

According to Uhl (2003), four principal Quaternary-Neocene aquifers are in the Afghanistan portion of the Kabul river basin. They include:

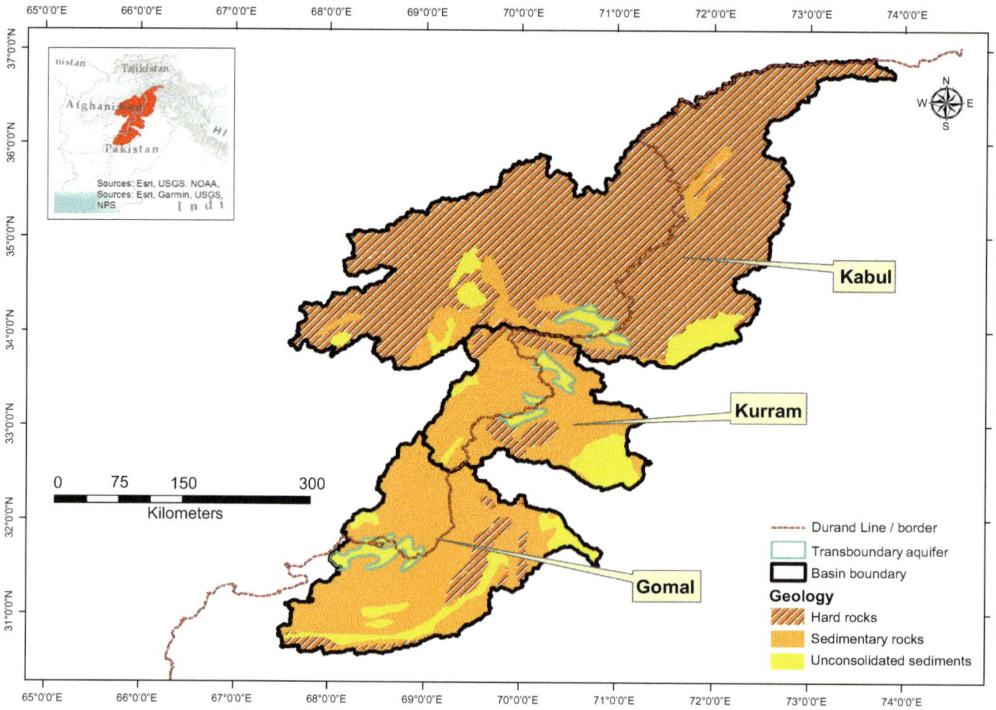

Fig. 6.2. Distribution of major aquifers in the Kabul, Kurram and Gomal river basins. (Adapted from Wandrey and Law, 1998)

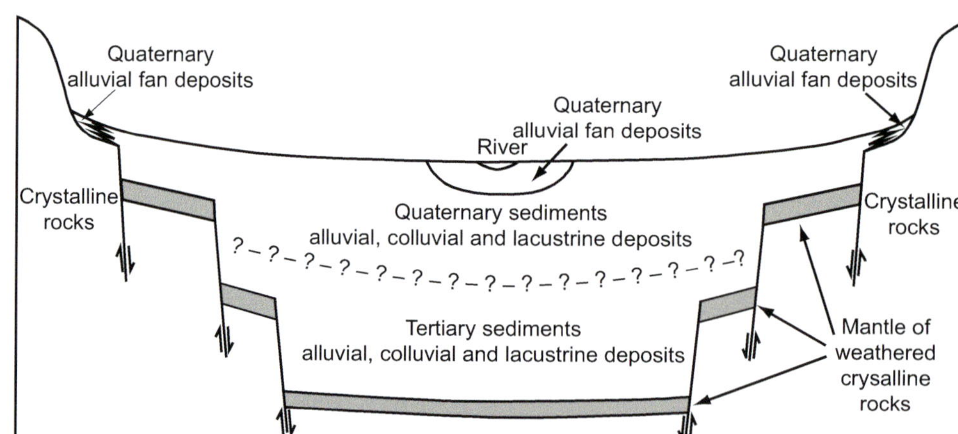

Fig. 6.3. Generalized section of the hydrogeological units in the Kabul river basin in Afghanistan. (Adapted from Broshears *et al.*, 2005)

- Logar basin to the south;
- Kabul river basin in the Kabul City area, including the Paghman area to the west;
- Parwan basin associated with the Shamali Plain to the north of Kabul and the Panjsher basin and its tributaries; and
- Jalalabad basin to the east.

These aquifers are broadly indicated in Fig. 6.4. High-resolution mapping of the Quaternary geology by Doebrich and Wahl (2006) suggests a highly complex distribution of aquifers across the basin controlled largely by tectonic activity.

Aquifer permeabilities are generally high, ranging from 2 m to 110 m/day. Transmissivity is highly variable and reportedly ranges from 10 m^2 to 8000 m^2/day. Alluvium has hydraulic conductivity ranging from 46 m to 105 m/day. Well yields may be as high as 70 L/sec (Broshears *et al.*, 2005).

Cementation of pore spaces by secondary mineralization of the deeper parts of the aquifer is expected, generally decreasing well yields with depth (Houben *et al.*, 2009). The deep Neocene aquifer is poorly recharged and contains essentially non-replenishable 'fossil' water. The development of this aquifer would likely be unsustainable (Saffi, 2019).

Aquifers are often covered by low-permeability loess deposits acting as an essential protective barrier (Houben *et al.*, 2009). Around river courses, loess layers are sometimes eroded. Intercalated loess and gravels in the upper parts of the aquifers can produce localized perched aquifers of low productivity. These shallow phreatic aquifers tend to be drawn upon by shallow wells but may quickly run dry during drought periods.

Groundwater flow is primarily through saturated alluvium and other basin in-fill deposits. The water table surface generally mirrors topographic patterns, and thus groundwater generally flows in the direction of surface-water discharge. Most apparently, groundwater flows from the western or south-western margins to the north-east through the very narrow gorge of Tangi Gharu (Houben *et al.*, 2009). Depressions without natural outlets in the centre of the basin

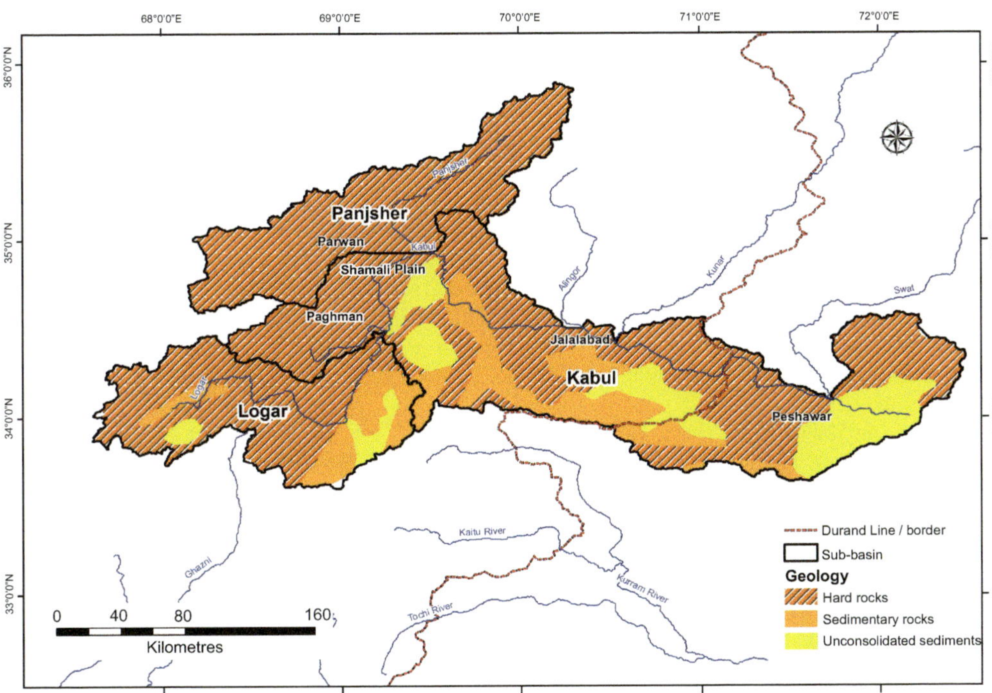

Fig. 6.4. General distribution of aquifers inside the Kabul river basin.

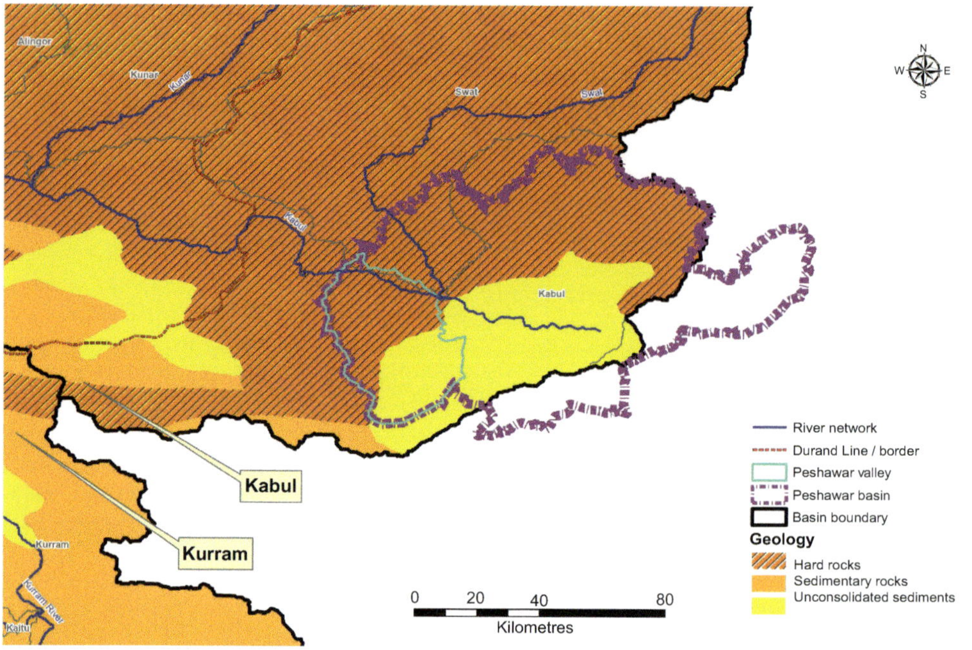

Fig. 6.5. Distribution of aquifers downstream of Warsak Dam in Peshawar basin and Peshawar Valley, Pakistan.

have created groundwater-fed saline swamps in areas.

Hydrogeologically, the unconsolidated sediments situated within the lower part of the Kabul river basin within Pakistan are broadly classified as either shallow (i.e. <60 m in depth) or deep (i.e. 60– 100 m in depth). Both aquifers exhibit heterogeneity both in the lateral and vertical extent (Kruseman and Naqavi, 1988). The shallow aquifer is unconfined with a hydraulic conductivity (K) range of 1.6–4.8 m/day. The deep aquifer has K values that range from 0.07 m to 0.3 m/day.

Along the Kabul river, numerous escarpments can be seen, the majority of which are composed of alluvial fan deposits. Additionally, the Peshawar basin contains horizontally bedded sand, silt and clay (Cornwell, 1998). The south-eastern section of Peshawar, reaching all the way to Nowshera (see Fig. 6.5), encompasses the middle flood plain near the Kabul river and is made up of sands and gravels characteristic of an alluvial fan system. Weathering and erosion have transported the sediments, implying a low degree of sorting in the sediments and a diverse composition of sand, gravel and pebbles associated with clays and other fine materials (Muhammad and Khalid, 2017).

The sediments in the shallow subsurface up to a depth of more than 5 m are a mixture of silty clay, clay and gravel. However, below 10 m and up to 60 m, it reveals a network of interconnected strata ranging in thickness from coarse to fine. Further down, to depths of up to 150 m, its composition is finer and made up of clay and silt mixed with sand and gravel.

Groundwater Recharge

In Afghanistan, the primary source for groundwater recharge to the unconsolidated Quaternary and Neocene aquifers is snowmelt, infiltrated through streambeds and banks and is driven by the flow season with a peak during the spring (Houben *et al.*, 2009; Sadid, 2020). Thus, if snow cover in the mountainous areas is limited, then the water availability in subsequent years will be limited, as was demonstrated by the 2000–2005 drought that significantly constrained recharge (Houben *et al.*, 2009).

Flow monitoring in the upper Kabul basin along the Maidan river between Gulbagh (upstream) and Chehlsutun (downstream) gauging stations indicates that 72% of the flow recharges the groundwater (Sadid, 2020). For the Logar river aquifer, 92% of recharge is derived from infiltration from rivers and channels, 4% from infiltration from basin margins at the interface between consolidated and unconsolidated sediments, and 4% directly from precipitation (Proctor and Redfern Int. Ltd, 1972). These studies clarify that recharge is highly variable, spatially and temporally.

Recharge rates have high uncertainty but are generally moderately high. Proctor and Redfern Int. Ltd (1972) found groundwater recharge rates for the Logar, Kabul and Paghman basins are estimated to be 2.16 m^3/sec, 0.82 m^3/sec and 0.48 m^3/sec, respectively (Table 6.1). Sadid (2020), using multiple methods and uncertainty analysis, estimated the recharge to be 0.35–1.9 m^3/sec for Logar, 0.59–2.94 m^3/sec for Kabul and 0.27–1.12 m^3/sec for Paghman. Groundwater-level fluctuations in wells situated close to rivers strongly correlate with the rivers' flow discharge variations and weakly correlate with precipitation rates (Sadid, 2020). Annual recharge to the aquifers in the Kabul City area is in the range of 250 mm/year, a very high estimate given the semi-arid environment that may be attributed to the contributions from upland areas delivered by the major rivers (Uhl, 2003).

In Pakistan, groundwater recharge in all three basins comprises these components:

- infiltration from surface-water sources;
- infiltration from nearby foothills;
- precipitation infiltration (direct recharge);
- infiltration from irrigation canals and agricultural fields; and
- seepage from unlined canals and streams.

The major source of groundwater recharge in these basins comes from infiltration from the major rivers like the Kabul, Kurram and Gomal (surface-water sources) (Houben *et al.*, 2009). The other major reason groundwater recharges from surface waters is that all three basins have snow accumulation during winter, which melts during summer and increases the groundwater accumulation.

Table 6.1. Recharge rates for three major aquifers in the Kabul region.

	Recharge rates (m^3/sec)	
Basin	Proctor and Redfern Int. Ltd, 1972	Sadid (2020)
Logar	2.16	0.35–1.9
Kabul	0.82	0.59–2.94
Paghman	0.48	0.27–1.12

In Peshawar district, some rainwater infiltrates into the soil while the rest travels downhill as runoff or sheet flow. Kruseman and Naqavi (1988) claim that rainfall recharge occurs at higher elevations near foothills. Groundwater recharge is greatly influenced by the Bara and Kabul rivers, which enter the district from the south-west and the east. The Swat and Indus rivers also recharge the Peshawar Valley aquifers. Over half of Peshawar is devoted to agriculture, with four major canals diverting water from the Kabul river to provide irrigation. Approximately 25–30% of the irrigation water in the Peshawar area is returned to the water table. In comparison, more than 4% of the annual precipitation of 420 mm is infiltrated and recharges the groundwater systems (Kruseman and Naqavi, 1988). Similarly, in the Pabbi region near Peshawar district, approximately 37 mm of groundwater recharge occurs each year (Masud *et al.*, 2013). That study also showed that recharge in Pabbi is mainly due to irrigation return and canal percolation. It is widely accepted that the Kabul river's hydraulic connection to the aquifer is sometimes responsible for recharging the aquifer (Urban Policy Unit, 2014).

In the Kabul, Kurram and Gomal river basins, different regions have different soil and recharge capacity layers, as shown in Table 6.2. Alluvial deposits, which include sand, silt, clay, gravel and boulders, are the most common lithologies. When it comes to fluvial sediments, most are deposited on hillsides and in valleys, while fine-grained sediments are found in the valleys' flood plain. Alluvium thickness ranges from a few metres near the foothills to more than 300 m in the middle of the aquifer. Further exploratory well-drilling, seismic survey or other groundwater inquiry can be conducted to

Table 6.2. Rainfall infiltration percentages for different lithological layers.

Lithology	Geology	Rainfall infiltration (%)
Alluvium	Unconsolidated sediments	5
Piedmont		10
Playas		15
Conglomerates	Sedimentary rocks	7
Limestone		8
Sandstones		6
Shales		1
Mudstones		2
Shale/limestone		4
Shale/sandstone		3
Intrusive bodies	Hard rocks	1

identify the exact depth to bedrock. The literature review, on the other hand, suggests that the depth of bedrock in these basins fluctuates and is particularly deep in the Peshawar Valley's core. With an average elevation of 300 m, the valley's topography ranges from flat to undulating, with a maximum elevation of 700 m.

Groundwater Use

Drinking and domestic water

Groundwater drawn from shallow hand-pumped wells provides the primary source of drinking water in the Kabul river basin (Banks and Soldal, 2002; Houben *et al.*, 2009). Around 80% of the population relies on shallow wells fitted with hand pumps, whereas only 20% has access to piped water. Survey data collected by UNCHS-Habitat in 2000–2001 shows that around 30–40% of wells in Kabul are dry or not functioning. Withdrawal rates have risen sharply due to population growth combined with internal migration. According to the current Afghanistan Water Law of 2009 (the Act; details are provided in Chapter 8), provision of water supply for drinking and livelihoods, including the construction of water-treatment plants, water-conveyance facilities, sewerage systems and sewage-treatment plants are of primary importance. The Act declares that 'use of water is free' but that water service providers may charge users a fee for the supply, storage, transmission, diversion, treatment, operation and maintenance of the water supply and irrigation systems.

Similarly for Pakistan, the Water and Power Development Authority (WAPDA) Act 1958 (see section 'Pakistan Water Laws and Institutions' in Chapter 8), grants WAPDA the authority to, among other things, design water-supply and drainage systems, as well as projects for irrigation and recreational water usage, flood control and inland navigation. The Act also grants WAPDA authority, subject to other laws, and jurisdiction over the groundwater resources of any territory within a province. However, before WAPDA exerts any control over groundwater resources, the government must notify WAPDA of the area over which, and to what extent, it has control. In KPK, groundwater from the Peshawar Valley aquifers serves the water needs of five of Pakistan's most populous districts (Peshawar, Mardan, Swabi, Charsadda and Nowshera), which have a combined population of over 11.4 million (37% of the total population of 30.5 million in the province).

Approximately 840,000 m^3/day of groundwater is abstracted from different sources like public wells, hand pumps and dug wells in the Peshawar basin (downstream of Warsak Dam). Khan *et al.* (2014) calculated the basin-wide annual average variation in groundwater storage as given in Table 6.3, and the overall groundwater abstraction wells present in Peshawar Valley are shown in Fig. 6.6. The details of these abstraction points against groundwater abstraction are presented in the table. The trend across the whole basin shows a significant decline in groundwater with a single sub-basin as an exception, which shows a slight increase in groundwater (Ali, 2018).

Approximately 670 government-operated tube wells and more than 1000 privately owned wells (including hand pumps and dug wells) are present in the Peshawar Valley. A survey conducted by Khan *et al.* (2014) showed that groundwater abstraction rates from these wells range from 2,900 to 387,000 m^3/day (Table 6.3). In order to meet their water needs in the face of climate change, farmers are turning to the High Efficiency Irrigation System (HEIS) (see Fig. 6.7).

Table 6.3. Groundwater abstraction details in the Peshawar Valley. (Adapted from Ali, 2018).

Abstraction point	Total number	Estimated abstraction* (m^3/day)
Public wells	670	387,000
Privately owned wells	552	2,900
Hand pumps	428	423,000
Dug wells	63	27,735

*Wells assumed to operate for periods ranging from 4 to 12 hours per day

Irrigation

For Afghanistan, the primary sources of cropland irrigation in the Kabul river basin are canals, streams, springs, karezes[3], dug wells and boreholes (Akhtar *et al.*, 2018). Nationally, at least 30% of all irrigation is derived from groundwater sources, divided equally between springs, karezes and shallow wells (Central Statistics Organization, 2018). Groundwater may be an important source of irrigation water in the Kabul basin provinces such as Kabul and Wardak, accounting for 33–42% of all sources

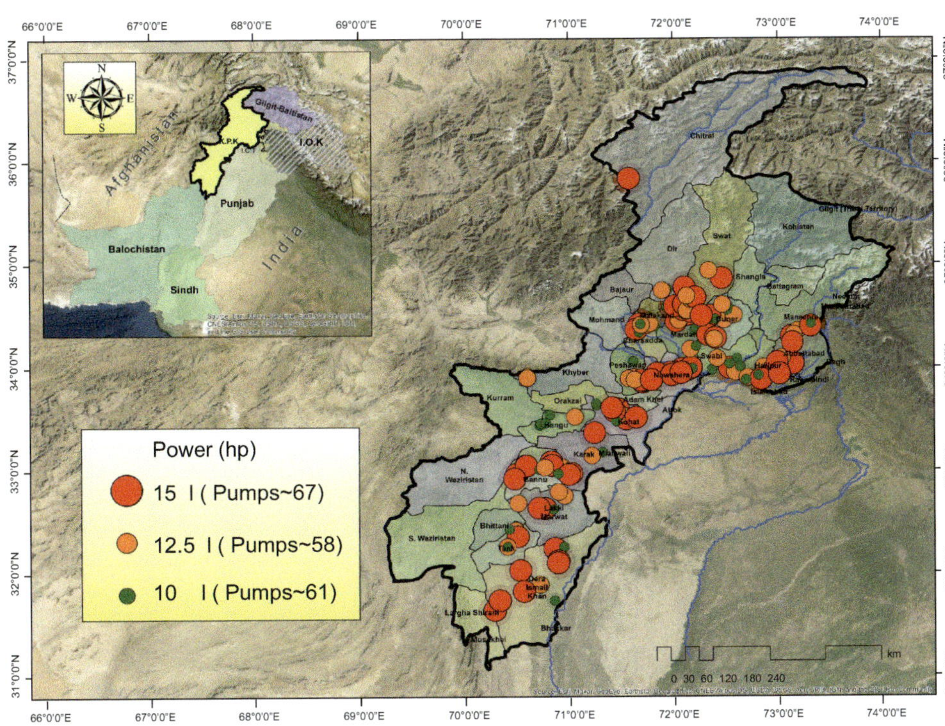

Fig. 6.6. Groundwater abstraction across the lower Kabul river basin for irrigation and non-irrigation wells. The installed pump capacity (hp) is indicated. (Adapted from Public Health Department and Agriculture Engineering Department)

Fig. 6.7. Installation of drip-irrigation system in command area of Gomal Zam Dam. (From: Kashif Hussain, IWMI-Pakistan)

(Shobair, 2001). Fig. 6.1 shows that in transboundary Gomal River Basin, due to high prices of fuel people are adopting solar pumps for the abstraction of groundwater for irrigation in the area.

Lift irrigation from dug wells and boreholes is a more recent irrigation method that is rapidly expanding. With the increasing frequency and severity of droughts in recent decades (Qutbudin *et al.*, 2019), the use of deeper groundwater, abstracted via pumped dug wells and boreholes, has increased rapidly. Private farmers, as well as NGOs, are initiating drilling.

For Pakistan, the communities in all three transboundary basins rely on groundwater sources to provide around 60% of irrigation supplies. Agriculture consumes more than 93% of water, although water-use efficiency is relatively low, ranging between 40% and 50% (Ahmad *et al.*, 2021). Groundwater has had a key influence in raising overall agricultural productivity (Jabeen *et al.*, 2021). Groundwater use in all three basins differs depending on its quality, depth and availability. In KPK, groundwater quality is generally good but due to aquifer characteristics and depths, its use for irrigation is limited compared to surface water.

Information on the irrigation schemes supplied with groundwater from wells is scant (Akhtar, 2017). It is known that individual farmers sometimes use shallow wells to irrigate their fields in the lower reaches of large canal irrigation schemes where water-shortage issues are common. Dwindling surface-water supplies because of falling storage capacity and mounting concerns about climate change are jeopardizing the future of irrigated agriculture, which produces almost 90% of Pakistan's grains (Basharat and Tariq, 2015). The arid and semi-arid conditions that predominate in most of the country necessitate irrigation for sustainable agricultural production, as evapotranspiration is high and rainfall is infrequent and unreliable (Khan *et al.*, 2017). Over 75% of this precipitation occurs during the monsoon season (July–September). Rain contributes approximately 30,000 Mm^3/year to agricultural irrigation, insufficiently covering 15% of crop-water requirements; the remaining 85% is satisfied through irrigation (Bhutta and Smedema, 2007). Due to the scarcity and instability of available surface water, farmers are forced to extract groundwater to meet irrigation water demands, with limited regard as to its quality or pumping cost.

Groundwater Management

Groundwater quality

Hydrochemically, the groundwaters of the Kabul basin in Afghanistan are classified as calcium bicarbonate type and generally high in hardness due to interactions with aquifer carbonates leading to near-neutral pH and high calcium and magnesium values (Afzali, 2017; Jawadi *et al.*, 2020). The redox conditions of the shallow groundwater are weakly toxic due to oxygen depletion, probably caused by the seepage of nutrient-rich sewage water. Redox and pH conditions limit trace metals and metalloids (Houben *et al.*, 2009). Water–aquifer interactions, combined with the arid climatic conditions, bring about elevated salinity and

Table 6.4. Summary of groundwater quality in the Kabul basin, Afghanistan. (Adapted from Afzali, 2017).

Parameter	Range
pH (-)	7.59–8.12
Turbidity (NTU)	2–4
TDS (mg/L)	288–681
Chloride (mg/L)	3.6–240
Sulphate (mg/L)	5–326
Bicarbonate (mg/L)	145–430
Carbonate (mg/L)	10–140
Calcium (mg/L)	23.6–531
Magnesium (mg/L)	9–142
Sodium (mg/L)	6–230
Boron (mg/L)	0.025–2.4
Fluoride (mg/L)	0.14–1.25
Ammonia (mg/L)	0.01–0.21
Nitrate* (mg/L)	3.1–54
Phosphate (mg/L)	0.05–1.05
Arsenic (mg/L)	0.002-3[2]
Chromium (mg/L)	0.01–0.03
Iron (mg/L)	0.04–1.7
Manganese (mg/L)	0.001–0.005
E. coli (cfu/100mL)	1.9–52.8

*Reported as nitrate, not nitrate-nitrogen.

undesirable constituents such as borate and iron (Table 6.4).

Overall, in all three basins, the groundwater quality is highly contaminated both microbiologically and chemically and does not always conform to the World Health Organization (WHO) drinking-water standards (see Figs 6.8 and 6.9). More than 60% of groundwater samples were analysed for important parameters that include total dissolved solids, total hardness, turbidity and magnesium. It was found that more than 90% of groundwater samples had magnesium levels that exceeded the standards set by the WHO (Jawadi *et al*., 2020). Similarly, it was identified by Afzali (2017) that 30% of groundwater samples were not suitable for drinking, with the highest microbial contamination found in urban areas, making the water unsafe for drinking without treatment (e.g. boiling). This analysis was based on samples collected from 50 boreholes within the Afghanistan portion of the Kabul river basin. Groundwater was suitable for irrigation in most cases. Comparison with historical data suggests that population increase and extended drought have had a detrimental effect on water quality. In some cases, long-term declines in groundwater levels are also resulting in water-quality deterioration (Saffi, 2019).

Microbial pollution of surface water throughout the entire Kabul river basin (Iqbal and Hofstra, 2018) likely also presents a significant

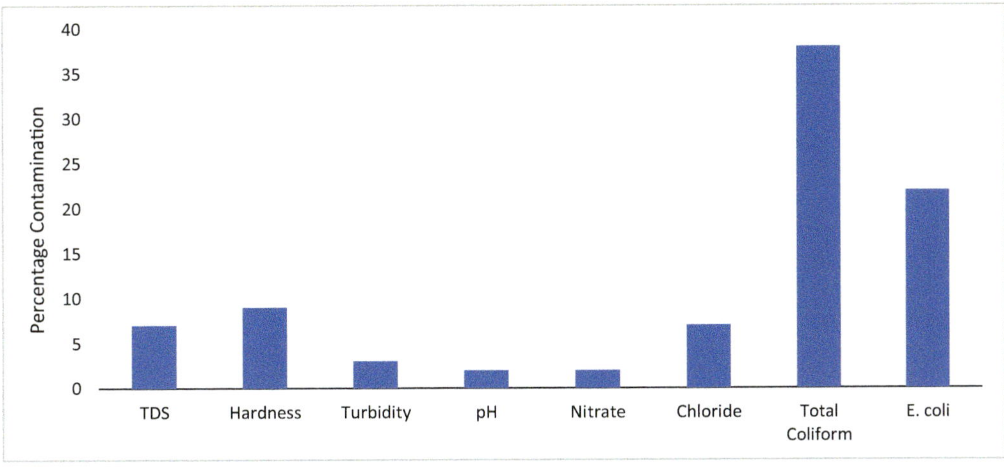

Fig. 6.8. Levels of contamination found in the groundwater samples. (Adapted from Imran *et al*., 2018)

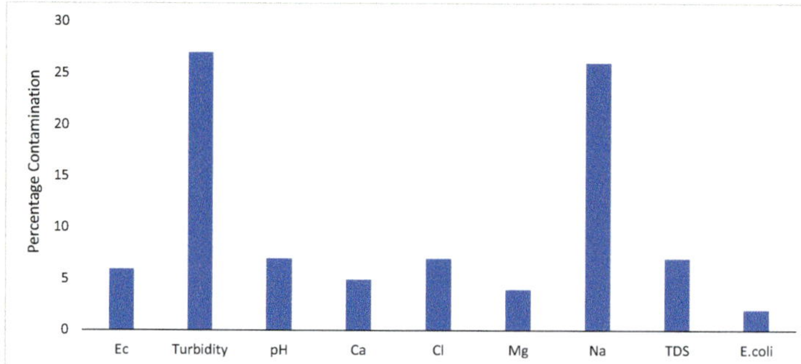

Fig. 6.9. Levels of contamination found in the irrigation water samples in the Gomal river basin and its tributaries. (Adapted from Muhammad and Ullah, 2021)

risk to shallow groundwater resources. The primary sources of pollution are leaky septic tanks and other sewerage infrastructure, drainage-pit latrines, roadside ditches, major rivers, irrigation channels, solid waste disposal and agrochemical use (Zaryab *et al.*, 2017). Houben *et al.* (2009) partly attribute high infant mortality to inadequate water hygiene. Commonly identified indicators of poor hygiene are elevated concentrations of nutrients such as nitrate and faecal bacteria. The high permeabilities of the sandy to gravelly Kabul basin aquifers offer no barrier to the spread of contaminants. The shallow loess beds do, however, provide a valuable barrier to the movement of microbial pollutants.

The water quality in the Peshawar basin has remained high over the past four decades, but increased urbanization and human activities have led to the detection of *E. coli* at various sites in the shallow aquifer (Muhammad and Khalid, 2017). Nasreen (2006) studied the soil, surface water and groundwater quality in Peshawar district to establish the state of environmental degradation. Physicochemical characteristics, including pH, electrical conductivity, temperature and total dissolved solids, were analysed. Groundwater and surface waters were also analysed for anions, cations and heavy metals. According to the findings, physical metrics were within acceptable ranges; however, anion and cation levels in the affected areas exceeded the standards. The movement of groundwater through the limestone, gypsum, dolomite, sulfide seams and halides were linked to those high concentrations.

Similarly, Adnan and Iqbal (2014) used a geographic information system (GIS) approach to map Peshawar district's groundwater quality. For determining the groundwater's physicochemical properties, samples were collected from 105 wells. Electrical conductivity, total dissolved solids, turbidity, pH, hardness, chloride and nitrate were among the variables examined. Physiochemical parameters were determined to be higher in urban areas. Some areas outside the towns have greater concentrations of both pH and nitrates. The water samples were analysed for 16 routine water-quality parameters as shown in Table 6.5.

A comprehensive investigation of the Kabul river's drinking-water quality and groundwater sources before its confluence with the Indus river (Imran *et al.*, 2018) shows that approximately 55% of the samples were judged safe for consumption relative to the WHO standards for drinking-water quality, as shown in Fig. 6.8.

A similar water-quality analysis has been performed by Muhammad and Ullah (2021) in the Gomal river basin and concluded that, except for turbidity and sodium, the water properties of the Gomal Zam Dam and its tributaries were within the WHO drinking-water requirements. According to the concentration values of parameters selected, Gomal Zam Dam and its tributary water are classified for irrigation and domestic use. The Gomal Zam Dam and its tributaries' irrigation water quality was classified as within allowed limits, as shown in Fig. 6.9.

For the 45% of samples judged unsafe, nearly all major districts in these basins had pH and turbidity levels that were outside the range

Table 6.5. WHO standards for drinking-water quality. (Adapted from World Health Organization, 2022)

Parameter	WHO standard	Parameter	WHO standard
Odour	Clear (unobjectionable)	Alkalinity as $CaCO_3$ (mg/L)	200–600
Turbidity (NTU)	5–10	Total hardness as $CaCO_3$ (mg/L)	300–600
Total dissolved solids (mg/L)	500–2000	Calcium as Ca (mg/L)	75–200
Electrical conductivity (μS/cm)	400	Magnesium as Mg (mg/L)	30–150
pH	6.5–8.5	Iron as Fe (mg/L)	0.3–1.0
Chromium as Cr (mg/L)	0.05	Nitrite as NO_2 (mg/L)	3
Nitrate as NO_3 (mg/L)	50	Fluoride as F (mg/L)	1–1.5
Chloride as Cl (mg/L)	250–1000	Sulphate as SO_4 (mg/L)	200–400
Chemical oxygen demand (mg/L)	10	Aluminium as Al (mg/L)	0.2

of recommended levels. Exceedances of nitrate, TDS and hardness were widespread in Peshawar, while the concentration of hardness, chloride and TDS in Banu was higher. In Nowshera, 9% of the water was contaminated with nitrate, along with sodium, chloride and TDS. Just 4% of samples from the Dara Ismail Khan area were found to have nitrate levels over permitted limits. Only pH was an issue in Peshawar district.

Overall, the surface and groundwater of the Kabul river are polluted. As per WHO standards, the water is not fit for drinking, bathing or cleaning. The highest amount of *E. coli* and other harmful bacteria is found in the surface and groundwater source of Kabul river in Peshawar Valley (Iqbal and Hofstra, 2018). Due to climate change, there will be more extreme events like floods that will further reduce water quality in the region. Similarly, Iqbal *et al.* (2018) found that the relationship between climatic variables with water quality, particularly with the sources of bacterial pollution, as well as temperature and precipitation, are directly correlated with bacterial concentration in water resources.

There is a strong relationship between socio-economic variables like population, livestock numbers, agriculture and climate change on the water-quality standards in the Kabul river basin (Iqbal *et al.*, 2019). Due to the increase in population and limited or restricted availability of sanitation practices in the Kabul river basin, all the untreated sewage and livestock manure is dumped or discharged into surface-water resources that contaminate the surface water and, ultimately, the groundwater. In the city of Nowshera, a sampling campaign was launched to monitor the water quality and found that drinking water is not fit for use due to the high amount of *E. coli* present in it (Javed *et al.*, 2014).

Groundwater quantity

The lack of regulation to control the construction of new wells has led to major impacts on the resource. Unchecked drilling of new wells and boreholes by farmers and NGOs has led to significant depletion of groundwater storage (UNAMA, 2016). Groundwater-level declines lead to the drying of shallow wells and kareze systems and decreased dry-season surface water availability (Saffi, 2019).

Groundwater levels have been declining in the 21st century. Monitoring 13 Ministry of Energy and Water (Afghanistan) observation wells between 2003 and 2018 revealed systematic declines in water levels ranging from 5 to 40 m (Saffi, 2019). The trends for monitoring wells within the Kabul river basin typically show a declining trend. Well MW_1 situated in District 12 of the Kabul City area, for example, shows groundwater levels have declined from 15.2 to 26.0 m (Fig. 6.10). Wells situated further from the stabilizing effect of interactions with rivers are known to have more pronounced signs of depletion (Aliyar, 2016).

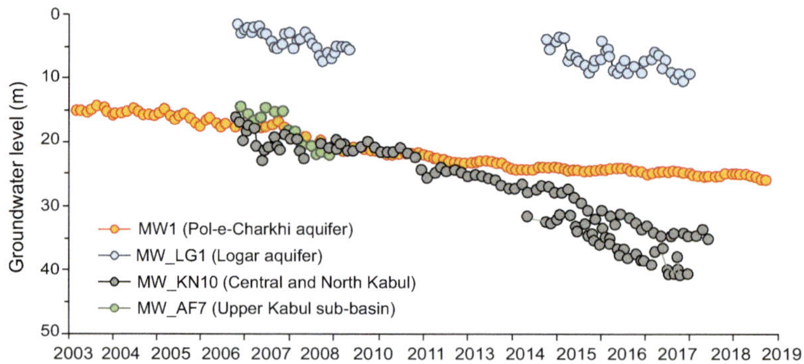

Fig. 6.10. Groundwater-level trend for four monitoring wells situated in different aquifers within the Kabul river basin in Afghanistan.

The increasing demand for groundwater resources, particularly for irrigation, is driven by recurring drought events combined with the growing availability of modern well drilling and pumping technologies (Akhtar, 2017). Increasing summer temperatures due to climate change are causing accelerated melting of the snowpack, adding further stress on groundwater resources (World Bank, 2010). Decreased groundwater availability could threaten the water supply of entire communities and lead to a range of humanitarian crises, including disease, population displacement and conflict (Savage *et al.*, 2009).

Managed aquifer recharge (MAR) has been proposed as an immediate solution for the drastic groundwater storage depletion in recent years (Noori and Nasimi, 2019; ADB, 2021). MAR relies on using 'surplus' surface water (rivers and snowmelt water), which would otherwise flow to downstream parts of the basin. Modelling results indicate that seasonal flow diversion from the Kabul river could increase groundwater storage by 9 Mm^3 (Masoom, 2018).

The hydrogeological conditions in the Peshawar Valley are complex, and examination of the aquifers in the Peshawar basin must be done with care. Recharge from yearly precipitation is 4%, or 17 mm, while recharge from surface-water irrigation losses is 92 mm each year. Forty per cent of the aquifer's water balance comes from the Kabul river, which helps offset the daily withdrawal of approximately 827,000 m^3 (Ali, 2018).

In Peshawar Valley, groundwater is mostly drawn from dug wells, springs, small boreholes and deep-tube wells. This groundwater is used for irrigation and domestic and industrial use in these areas. The irrigation system, dams and rivers that run through the area contribute significantly to aquifer replenishment. According to this, the aquifer's pumping depends on river recharge generated by the surface–groundwater interactions. The depth of the shallow-water table aquifer varies from 1.5 to 91 m deep. Except in a few locations in urban areas, pumping appears to be substantially less than recharge (Khan *et al.*, 2014).

For example, climate change could significantly impact Peshawar district's groundwater sustainability if the river experiences low flow. The flow direction of water is from the west and south to the east, towards the continuous head, at the core of the urban area, where the highest population density means pumping is at its highest concentration.

Conclusions

Groundwater resources of the three transboundary basins have high potential in areas with favourable hydrogeological characteristics and represent a vital resource for domestic and agricultural use. Due to unregulated development and climate change, these resources are being depleted. The trends are most pronounced in the most populous parts of the basin. Therefore, considering the depletion in groundwater storage, the current practices seem unsustainable over the long run and will have serious consequences if investment in sustainable groundwater management planning are not made. Such planning would need to consider environmental, social and economic aspects. Given the critical gaps in data, information and knowledge that have been revealed through this study, a number of key steps can be taken to promote sustainable management:

- Increase investments in gathering additional data to better understand the complex hydrogeology and hydrogeochemical environments. This is particularly needed for the Kurram, and Gomal river basins where information is most lacking.
- Establish basin-wide groundwater level and quality monitoring strategies that make data accessible to stakeholders to improve transparency and public awareness.
- Target research on the impacts of climate change on groundwater storage and on the broader opportunities for implementing managed aquifer recharge, among numerous other topics; this is needed to enhance the knowledge base that underpins the effective planning, development and management of groundwater resources.
- Given the highly interconnected nature of the relations between surface water and groundwater resources, adopt a holistic approach that encompasses all water resources like surface water, groundwater, environmental aspects and other water-management strategies.

Notes

[1] The South Asia region includes Afghanistan and Pakistan, together with Bangladesh, Bhutan, India, Maldives, Nepal and Sri Lanka.

[2] See:https://climate-diplomacy.org/sites/default/files/2020-10/Climate%20Chang%20in%20Afghanistan.pdf. (accessed 13 July 2022)

[3] Kareze: an underground channel through which water flows from an aquifer within mountains to open fields.

Acknowledgements

We wish to thank Ms Sumble Ghani and Dr Hammad Gilani from the IMWI Pakistan office for helping to prepare some of the basin-level maps and figures. We also would like to extend our thanks to Mr Waqas Ahmad for helping to collect information regarding literature on groundwater in Pakistan.

References

ADB (Asian Development Bank) (2021) Afghanistan: preparing the Kabul managed aquifer recharge project. Available at: https://www.adb.org/projects/48440-002/main (accessed 24 July 2022).

Adnan, S. and Iqbal, J. (2014) Spatial analysis of the groundwater quality in the Peshawar District, Pakistan. *Procedia Engineering* 70, 14–22. DOI: 10.1016/j.proeng.2014.02.003.

Afzali, A.O. (2017) Physicochemical and bacteriological quality of groundwater in Kabul basin, Afghanistan. *IOSR Journal of Agriculture and Veterinary Science* 10(03), 58–61. DOI: 10.9790/2380-1003025861.

Ahmad, H.S., Imran, M., Ahmad, F., Rukh, S., Ikram, R.M. *et al.* (2021) Improving water use efficiency through reduced irrigation for sustainable cotton production. *Sustainability* 13(7), 4044. DOI: 10.3390/su13074044.

Akhtar, F. (2017) *Water availability and demand analysis in the Kabul River Basin, Afghanistan.* Dissertation, Center for Development Research ZEF, University of Bonn, Germany.

Akhtar, F., Awan, U., Tischbein, B. and Liaqat, U. (2018) Assessment of irrigation performance in large river basins under data scarce environment—a case of Kabul River basin, Afghanistan. *Remote Sensing* 10(6), 972. DOI: 10.3390/rs10060972.

Ali, N. (2018) Groundwater assessment of the Peshawar District and its potential for future demand. Thesis, Flinders University, Adelaide, Australia.

Aliyar, A. (2016) Impact of climate change on groundwater in Kabul basin, Afghanistan. Master's thesis, Graduate School of Agriculture, University of Miyazaki, Japan.

Banks, D. and Soldal, O. (2002) Towards a policy for sustainable use of groundwater by non-governmental organisations in Afghanistan. *Hydrogeology Journal* 10(3), 377–392. DOI: 10.1007/s10040-002-0203-y.

Basharat, M. and Tariq, A.-R. (2015) Groundwater modelling for need assessment of command scale conjunctive water use for addressing the exacerbating irrigation cost inequities in LBDC irrigation system, Punjab, Pakistan. *Sustainable Water Resources Management* 1(1), 41–55. DOI: 10.1007/s40899-015-0002-y.

Bhutta, M.N. and Smedema, L.K. (2007) One hundred years of waterlogging and salinity control in the Indus valley, Pakistan: a historical review. *Irrigation and Drainage* 56(S1), S81–S90. DOI: 10.1002/ird.333.

Broshears, R.E., Chornack, M.P., Mueller, D.K. and Ruddy, B.C. (2005) Inventory of groundwater resources in the Kabul Basin, Afghanistan: *U.S. Geological Survey Scientific Investigations Report 2005-5090*. Reston, Virginia.

CDKN (Climate and Development Knowledge Network) (2014) *The IPCC's fifth assessment report: What's in it for South Asia?*. Overseas Development Institute and Climate and Development Knowledge Network. London.

Central Statistics Organization (2018) *Afghanistan Living Conditions Survey 2016–2017*. Kabul.

Cheema, M.J.M., Immerzeel, W.W. and Bastiaanssen, W.G.M. (2014) Spatial quantification of groundwater abstraction in the irrigated Indus basin. *Ground Water* 52(1), 25–36. DOI: 10.1111/gwat.12027.

Cornwell, K. (1998) Quaternary break-out flood sediments in the Peshawar basin of northern Pakistan. *Geomorphology* 25(3–4), 225–248. DOI: 10.1016/S0169-555X(98)00061-0.

Doebrich, J.L. and Wahl, R.R. (2006) "*Geologic and mineral resource map of Afghanistan: U.S. Geological Survey Open-File Report 2006–1038*" DOI: 10.3133/ofr20061038.

GGP (Groundwater Governance Project) (2016) Global framework for action to achieve the vision on groundwater governance. Groundwater governance: a global framework for action, GEF (Global Environment Facility). UNESCO-IHP (International Hydrological Program), FAO (Food and Agriculture Organization of the United Nations), World Bank and IAH (International Association of Hydrogeologists). Available at: http://www.groundwatergovernance.org (accessed 24 July 2022).

Hirji, R., Mandal, S. and Pangare, G. (eds) (2017) *South Asia Groundwater Forum: Regional Challenges and Opportunities for Building Drought and Climate Resilience for Farmers, Cities, and Villages*. Academic Foundation, New Delhi.

Houben, G., Niard, N., Tünnermeier, T. and Himmelsbach, T. (2009) Hydrogeology of the Kabul Basin (Afghanistan), part I: aquifers and hydrology. *Hydrogeology Journal* 17(3), 665–677. DOI: 10.1007/s10040-008-0377-z.

Imran, S., Bukhari, L.N. and Gul, S. (2018) *Water Quality Assessment Report Along the Banks of River Kabul Khyber Pakhtunkhwa 2018*. Pakistan Council of Research in Water Resources (PCRWR).

Iqbal, M.S. and Hofstra, N. (2018) Modeling *Escherichia coli* fate and transport in the Kabul River Basin using SWAT. *Human and Ecological Risk Assessment* 25(5), 1279–1297. DOI: 10.1080/10807039.2018.1487276.

Iqbal, M.S., Dahri, Z.H., Querner, E.P., Khan, A. and Hofstra, N. (2018) Impact of climate change on flood frequency and intensity in the Kabul River Basin. *Geosciences* 8(4), 114. DOI: 10.3390/geosciences8040114.

Iqbal, M.S., Islam, M.M.M. and Hofstra, N. (2019) The impact of socio-economic development and climate change on *E. coli* loads and concentrations in Kabul River, Pakistan. *Science of the Total Environment* 650, 1935–1943. DOI: 10.1016/j.scitotenv.2018.09.347.

Jabeen, M., Rashid Ahmed, S. and Ahmed, M. (2021) Enhancing water use efficiency and grain yield of wheat by optimizing irrigation supply in arid and semi-arid regions of Pakistan. *Saudi Journal of Biological Sciences* 29(2), 878–885. DOI: 10.1016/j.sjbs.2021.10.018.

Javed, F., Ahmed, M.N., Shah, H.U., Iqbal, M.S., Wahid, A. and Ahmad, S.S. (2014) Effects of seasonal variations on physicochemical properties and concentrations of fecal coliform in river Kabul. World Applied Sciences Journal 29(1), 142-149.

Jawadi, H.A., Sagin, J. and Snow, D.D. (2020) A detailed assessment of groundwater quality in the Kabul Basin, Afghanistan, and suitability for future development. *Water* 12(10), 2890. DOI: 10.3390/w12102890.

Khan, G.D., Ali, M. and Akbar, F. (2014) Over exploitation of groundwater resources and their influence on groundwater in Peshawar Valley. Civil and Environmental Research 6(4), 85–90.

Khan, H.F., Yang, Y.C.E., Ringler, C., Wi, S., Cheema, M.J.M. *et al.* (2017) Guiding groundwater policy in the Indus Basin of Pakistan using a physically based groundwater model. *Journal of Water Resources Planning and Management* 143(3). DOI: 10.1061/(ASCE)WR.1943-5452.0000733.

Kruseman, G.P. and Naqavi, S.A.H. (1988) Hydrogeology and groundwater resources of the North-West Frontier Province Pakistan. In: *Hydrogeology and Groundwater Resources of the North-West Frontier Province Pakistan*. Water and Power Development Authority, Hydrogeology Directorate, TNO Institute of Applied Geoscience.

Mack, T.J., Chornack, M.P. and Taher, M.R. (2013) Groundwater-level trends and implications for sustainable water use in the Kabul Basin, Afghanistan. *Environment Systems and Decisions* 33(3), 457–467. DOI: 10.1007/s10669-013-9455-4.

Masoom, M.F. (2018) Artificial recharge of groundwater as a management tool for the Kabul Basin, Afghanistan. Louisiana State University Master's thesis 4653. Available at: https://digitalcommons.lsu.edu/gradschool_theses/4653 (accessed 22 September 2022).

Masud, T., Ali, A., Hussain, H., Shahzada, K., Alam, B., Masud, N. and Rizwan, M. (2013) Estimation of groundwater balance for the Pabbi region, Khyber Pakhtunkhwa. Journal of Himalayan Earth Sciences 46(2), 113–119.

Muhammad, S. and Khalid, P. (2017) Hydrogeophysical investigations for assessing the groundwater potential in part of the Peshawar basin, Pakistan. *Environmental Earth Sciences* 76(14), 494. DOI: 10.1007/s12665-017-6833-0.

Muhammad, S. and Ullah, I. (2021) Spatial and seasonal variation for water quality indices of Gomal Zam Dam and its tributaries in the South Waziristan district, Pakistan. *In Review*. DOI: 10.21203/rs.3.rs-795754/v1.

Nasreen, S. (2006) Monitoring of surface water, ground water, air and soil in Peshawar Basin against time: the 3rd dimension. University of Peshawar, Pakistan.

NDMC (National Drought Monitoring Centre) (2014) Drought Bulletin 2014. Pakistan Meteorological Department. Available at: https://www.ndmc.pmd.gov.pk/quater414.pdf (accessed 3 June 2022).

Noori, K.M.A. and Nasimi, M.N. (2019) Kabul city groundwater and need for artificial recharge. Proceedings of the 4th International Conference on Civil, Structural and Transportation Engineering (ICCSTE '19), Ottawa. Paper no. ICCSTE 215. DOI: 10.11159/iccste19.215.

Proctor and Redfern Int. Ltd (1972) *Water Supply Sewerage Drainage and Solid Waste Systems for Greater Kabul: Joint Interim Master Plan*. Report for Royal Government of Afghanistan.

Qureshi, A.S. (2011) Water management in the indus basin in pakistan: challenges and opportunities. *Mountain Research and Development* 31, 252–260. DOI: 10.1659/MRD-JOURNAL-D-11-00019.1.

Qutbudin, I., Shiru, M.S., Sharafati, A., Ahmed, K., Al-Ansari, N. *et al.* (2019) Seasonal drought pattern changes due to climate variability: case study in Afghanistan. *Water* 11(5), 1096. DOI: 10.3390/w11051096.

Sadid, N. (2020) Surface-groundwater interaction in the Kabul region basin. Afghanistan Research and Evaluation Unit, Issues Paper, March. Available at: https://www.researchgate.net/publication/342068342_Surface-groundwater_interaction_in_the_Kabul_region_basin (accessed 8 March 2023).

Saffi, M.H. (2019) National is alarming on groundwater natural storage depletion and water quality deterioration of Kabul City and immediate response to the drinking water crises. DACAAR, Kabul. Available at: https://www.researchgate.net/publication/342068342_Surfacegroundwater_interaction_in_the_Kabul_region_basin (accessed 8 March 2023).

Satoh, Y., Kahil, T., Byers, E., Burek, P., Fischer, G. *et al.* (2017) Multi-model and multi-scenario assessments of Asian water futures: The Water Futures and Solutions (WFaS) initiative. *Earth's Future* 5(7), 823–852. DOI: 10.1002/2016EF000503.

Savage, M., Dougherty, B., Hamza, M., Butterfield, R. and Bharwani, S. (2009) Socio-economic impacts of climate change in Afghanistan: a report to the Department for International Development. Stockholm Environment Institute. Available at: https://www.weadapt.org/sites/weadapt.org/files/legacy-new/placemarks/files/5345354491559sei-dfid-afghanistan-report-1-.pdf

Shobair, S.S. (2001) Effects of deep wells on groundwater resources in Afghanistan. Unpublished paper, reference AFG/96/004. In: *Food and Agriculture Organization of the United Nations (FAO)*. Peshawar, Pakistan.
Uhl, V.W. (2003) Afghanistan: an overview of groundwater resources and challenges. Uhl, Baron, Rana Associates, Inc., Washington Crossing, Pennsylvania.
UNAMA (United Nations Assistance Mission in Afghanistan) (2016) Water rights: an assessment of Afghanistan's legal framework governing water for agriculture. Rule of Law Unit, October. Available at: https://unama.unmissions.org/sites/default/files/2016_19_10_water_rights_final_v2.pdf (accessed 8 March 2023).
Urban Policy (2014) *Master Plan Volume 1: Water Supply, Treatment and Distribution*. Planning and Development Department, Government of Khyber Pakhtunkhwa. Available at: http://urbanpolicyunit.gkp.pk/wp-content/uploads/2018/02/Water-Sanitation-Peshawar-Report.pdf (accessed September 2022).
Wandrey, C.J. and Law, B.E. (1998) "Maps showing geology, oil and gas fields and geologic provinces of South Asia. U.S Geological Survey Open-File Report 97-470C." DOI: 10.3133/ofr97470C.
World Bank (2010) *Scoping Strategic Options for the Kabul River Basin Development: A Multisectoral Decision Support System Approach*. Washington, DC.
World Bank (2020) *Managing Groundwater for Drought Resilience in South Asia*. Washington, DC. DOI: 10.1596/33332.
World Bank (2021) *Managing Groundwater Resources in Pakistan's Indus Basin: Present and Future Prospects*. Washington, DC.
World Health Organization (2022) *Guidelines for Drinking-Water Quality:* 4th edn incorporating the first and second addenda. WHO, Geneva. Licence: CC BY-NC-SA 3.0 IGO.
Zaryab, A., Noori, A.R., Wegerich, K. and Kløve, B. (2017) Assessment of water quality and quantity trends in Kabul aquifers with an outline for future drinking water supplies. *Central Asian Journal of Water Research* 3(2), 3–11.

7 Land and Water Use

Karthikeyan Matheswaran and Taimoor Akhtar

Chapter Overview

Two remote-sensing datasets were used to estimate land and water use in the Kabul, Kurram and Gomal transboundary basins shared between Afghanistan and Pakistan. The proportion of different land-cover classes within these three basins was estimated. Barren land and rangeland form the largest block of land-cover classes owing to the prevailing semi-arid conditions. Domestic water use ranges from 21.2 million m^3 (mcm) in Gomal to 554 mcm in Kabul and 106 mcm in the Kurram. Due to the lack of data, industrial water use was assumed to be one quarter of domestic water use. According to the cropland area estimated using the Regional Land Cover Monitoring System (RCLMS) and evapotranspiration data, agricultural water use is 7350 mcm, 1890 mcm and 716 mcm in the Kabul, Kurram and Gomal basins, respectively. Improving transboundary co-operation can help optimize land and water use in the basins. Given the difficult context in which water co-operation would start from zero, 'low-hanging fruit' for advancing co-operation may lie in: (i) focusing effort on the Kurram and Gomal where, given land- and water use patterns, efforts towards co-operation may be less contentious; and (ii) focusing on data collection and assessment to establish an evidence-based foundation on which co-operation can build.

Fig. 7.1. Gomal riverbed, Dera Ismail Khan district. (From: Kashif Hussain, IWMI-Pakistan)

DOI: 10.1079/9781800622371.0007

Introduction

Agriculture remains the key driver of land and water use and the main source of livelihoods for people in the Kabul, Kurram and Gomal basins. People living here depend on irrigation to a large extent for the cultivation of their lands due to the prevailing semi-arid and arid conditions (Fig. 7.1). Rainfall in many parts of the region is sufficient to meet only 15% of crop water requirements and irrigation is the mainstay of agricultural activities (Qureshi *et al.*, 2010). In Afghanistan, the agriculture sector employs 80% of the country's population and accounts for 90% of the water use (UNAMA, 2016). In Pakistan, 62% of the population depends on agriculture, which accounts for 95% of the water use (Qureshi and Ashraf, 2019). Farmers in the region continue to suffer from highly variable flows during the growing season because of rainfall variability and droughts.

The agricultural sector forms the backbone of livelihoods in the Afghanistan–Pakistan region and contributes around 70–75% of export revenue through agri-based textiles and agrifood products (TDAP, 2016; WTO, 2020). Climate change is likely to exacerbate changes in rainfall patterns, heat stress at critical phenological stages, and increased water requirements of crops (Waqas *et al.*, 2020; Syed *et al.*, 2022). Productive and sustainable use of land and water resources within these basins will be key to addressing food, nutrition, livelihood security and climate change challenges.

Land use and water resources are intrinsically interconnected. Spatio-temporal heterogeneity in land characteristics heavily influences the hydrological process which determines the quantity and quality of water available in a river basin (Weatherhead and Howden, 2009). The challenges in resolving issues with land and water management are inextricably linked with progress towards the United Nations Sustainable Development Goals (SDGs). The Gomal, Kabul and Kurram together cover 157,619 km^2 or 10.3% of the total area of Afghanistan and Pakistan, and 13.6% of the Indus river basin. The three basins are considered global hunger and severe malnutrition hotspots (von Grebmer *et al.*, 2021). Despite the significant potential in these basins, myriad land and water challenges confront different uses and users. Access to safe and secure water resources for drinking and sanitation remains a major challenge, which is evident from the low percentage of the population with access to safe and affordable drinking water in both Afghanistan (28%) and Pakistan (37%) (WHO and UNICE, 2021).

Water management challenges confronting upstream Afghanistan sections of the basins are markedly different from downstream basin areas in Pakistan. In the Afghan portion, water resource challenges for agriculture manifest in different ways due to poor irrigation infrastructure, insufficient suitable land, and constrained access to irrigation technologies, and coupled with a lack of institutional support, these have consistently dampened crop yields (Saleem and Raouf, 2011). War-torn Afghanistan has witnessed the breakdown of traditional social institutions like farmer-managed traditional irrigation systems resulting in severe damage to the agrarian community (World Bank, 2010). Despite some success in the past decade, investments to rehabilitate traditional irrigation systems or construct new infrastructure remain inadequate. The deepening economic crisis as a result of recent political change in Afghanistan may reduce the financial resources available for the rehabilitation of irrigation infrastructure. In contrast, the downstream parts of these three basins in Pakistan host a large number of urban and rural settlements and an extensive, well-developed irrigation network which has enabled increased crop production. Major challenges in the Pakistani portions of the basins include water-logging and salinity, unsustainable groundwater use, low water and land productivity on farms, inequitable water distribution, unreliable water delivery, and insufficient cost recovery (Faruqee, 1997; Qureshi *et al.*, 2004).

Afghanistan and Pakistan need to make rapid strides to achieve universal and equitable access to safe and affordable drinking water for all (SDG 6.1) by 2030 since only 28% of the Afghan population and 37% of the Pakistani population have access to safe drinking water (WHO and UNICE, 2021). To improve economic and food security, both countries plan to expand the irrigated area to meet their food production needs. In addition, both countries plan to tap hydropower potential to support energy security (Danish *et al.*, 2017; Sibtain *et al.*, 2021). The development and management of water resources in the region will have inter-sectoral implications on food and energy security. These challenges and trade-offs are leading to the application of integrated methods such as the water–energy–food nexus approach for finding synergies that reduce loss, increase efficiency, and find new resources (Wescoat *et al.*, 2021). Increasing the land and water productivity of irrigated areas in Pakistan offers substantial prospects to produce more food and ensure productive water and energy use. Knowledge of current land and water uses is imperative for developing sustainable strategies geared to meet food production, and economic development to meet the needs of rising populations, while addressing the twin challenges of biodiversity decline and climate change impacts (Archer *et al.*, 2010).

In this chapter, the authors offer a quantitative estimate of the state of land and water use in the Gomal, Kabul and Kurram basins for 2018 by leveraging the availability of regional and global remote-sensing data sources. Where possible, information from remote-sensing data was combined with secondary data and published sources to provide insight into land use. Following a review of data used for this effort, land use and land-cover estimates are first presented. Water use is then examined in the context of two sectors: agriculture and domestic. The chapter concludes with reflections on challenges and opportunities for integrated basin management.

Data Collection and Processing

Accurate and timely land and water use data are critical for agricultural planning, monitoring biodiversity hotspots, environmental provision and urban infrastructure planning. Monitoring land and water use of river basins can help identify subsistence farming areas, particularly in mountain regions, that could be transformed into market-oriented agricultural production systems (Sharma *et al.*, 2016). However, in developing countries, the data needed for resource management are often unavailable at the required spatio-temporal scale due to limitations in technical capacity and institutional challenges (Saah *et al.*, 2019). While the agricultural area under production is often available at a national and provincial scale, it is not adequate for basin-level studies due to scale issues and transboundary extents. For instance, in Afghanistan, national land-cover assessments were conducted with different types of classification systems in 1972, 1993 and 2010 using satellite data from various sensors. Since then, no national land-cover assessments have been conducted with a harmonized or standardized land-cover classification system. Remote sensing and actual evapotranspiration (ETa) data remain the main sources of information in data-scarce regions like Afghanistan. Depending on the purpose, these freely available products can be developed on a global to local scale. When basin-level estimates of land and water use are not available, global and regional products derived from remote-sensing data serve as an alternative to field data encompassing the whole of transboundary basins.

Due to the lack of basin-specific data on land and water use covering the Kabul, Kurram and Gomal river basins, the authors relied on two remote-sensing datasets, namely the SERVIR-HKH Regional Land Cover Monitoring System (RCLMS), available at 30 m spatial resolution for determining the state of LULC in target basins (Uddin *et al.*, 2021) and Synthesized Global ETa at 1 km spatial resolution for estimating agricultural water use (Elnashar *et al.*, 2021). For determining domestic water use, gridded population data from Worldpop (TDAP, 2016) were used in conjunction with an average per capita domestic water requirement of 55 litres per capita per day (lpcd) for the Kabul basin (Frischmann, 2012). Industrial water use is estimated to be 25% of the domestic water supply requirements (USAID, 2014).

The SERVIR-HKH RLCMS produced annual land-cover maps for the entire Hindu kush Himalayan region using a harmonized classification scheme for the period from 2000 to 2018 (Uddin *et al.*, 2021). The RLCMS used land-cover primitives and supervised machine-learning algorithms in the Google Earth platform to produce land-cover data suitable for the regional level. The methodology used over 51,000 sampling points to classify and validate the nine land-cover classes achieving an overall accuracy of 81.7%. This regional remote-sensing LULC dataset encompasses all three basins under consideration and is deemed adequate for assessing the state of land cover within the target basins. This study uses the 2018 dataset to assess land-cover patterns in these three target basins. Global actual evapotranspiration (ETa) datasets were retrieved from the global ETa dataset data available at 1 km resolution and monthly temporal resolution (Elnashar *et al.*, 2021). Better-resolution (500 m pixel size and eight-day repeat cycle) Moderate Resolution Imaging Spectroradiometer (MODIS) ETa data based on the Penman-Monteith equation are available but tend to underestimate high ET values and overestimate low values (Mu *et al.*, 2011). Its direct application for croplands in arid and semi-arid regions of Afghanistan and Pakistan is yet to be evaluated. ETa data for croplands in the three basins were estimated by overlaying the Synthesized Global ETa dataset onto the agricultural land-cover area delineated by the RLCMS dataset.

Land Use and Land Cover

Kabul river basin

The Kabul river basin covers an area of 92,000 km^2 of which 57.9% is in Afghanistan and 42.1% in Pakistan (Fig. 7.2). Rangelands cover around 54.81% of the basin area followed by 18.48% of barren land. Snow and glacial melt are mostly concentrated in the north-east part of the catchment (covering 6.81% of the basin area) and are the dominant sources of water that make up the Kabul river discharge. Forest covers 5.39% of the basin area and is found predominantly in the north-eastern part (93% of Afghanistan's forest area including the Nuristan Forest National Reserve). The Kabul river basin supports numerous densely populated urban and rural settlements. Occupying approximately

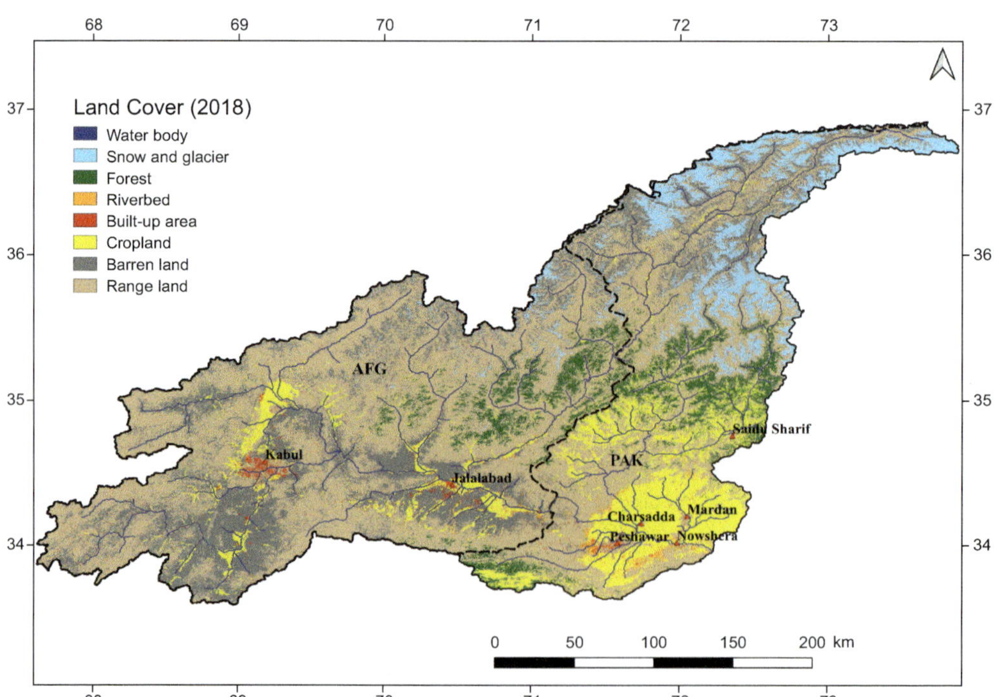

Fig. 7.2. Kabul river basin land cover 2018.

1% of the basin area, built-up LULC comprises large towns and cities (Kabul, Asadabad and Jalalabad in Afghanistan, and Peshawar, Nowshera, Charsadda and Mardan in Pakistan).

Cropland comprises 12.64% of the basin area covering 1.43 M ha.[1] The population in the Afghan portion of the basin is concentrated along river channels and valleys to leverage water availability for the summer crop. Large agricultural areas in the upstream portion of the basin include a broad plain stretching southward from the Ghorband and Panjshir rivers, the lower Logar Valley, areas adjacent to Kabul, and the wide valley of the Kabul river east of Jalalabad (World Bank, 2010). When it comes to downstream Pakistan, the districts of Charsadda, Mardan and Peshawar form a large contiguous extent of cropland. In addition, Bajaur, Dir, Swat and Khyber districts contain significant agricultural areas as a portion of the Kabul river basin. Major reservoirs in the Kabul river basin include Naghlu, Sarobi and Darunta located on the Afghanistan portion of the Kabul sub-basin.

About 80% of the installed hydropower capacity in Afghanistan is located in this basin. There are several large and small dams also in the downstream Pakistan section with proposals to add more large dams on the Swat, Panjkora and Kabul rivers (Taraky *et al.*, 2021). Over 80% of precipitation occurs as snow, mainly during the winter in areas with an elevation greater than 2500 masl (Rout, 2008). Large variations in precipitation between upstream highlands and the rest of the basin render the timing and duration of snowmelt crucial for water availability in the agricultural areas downstream.

Of the six sub-basins in Table 7.1, only Kunar and Kabul are shared by Afghanistan and Pakistan. Panjsher, Alingor and Logar are completely within Afghanistan while the Swat sub-basin is entirely within Pakistan (Fig. 7.3). Kabul sub-basin has large urban settlements including Kabul, Jalalabad and Peshawar, and has the largest percentage of built-up area among the six sub-basins (2.35%). Snow and glacier area form a significant part of the Kunar and Swat sub-basins with 17.88% and 9.67% of the basin area, respectively. An inventory of glaciers within the Kabul basin estimated 1600 glaciers with the highest concentration in the Kunar and Swat sub-basins (Williams, 2013). The snow and glacier meltwater from these two basins contribute significantly to the discharge into the Kabul and Indus river basins, particularly during the summer months. Both the Swat and Kabul sub-basins host many reservoirs, which are critical for domestic water supply, irrigation and hydropower production. The Swat and Kabul sub-basins comprise the highest cropland percentage (~24% of basin area). The remaining sub-basins consist of croplands between 2.5% and 6% of the basin area. The cultivated area in the Kabul basin is projected to increase with the expected population growth in both countries (Najmuddin *et al.*, 2017; Anwar and Bhatti, 2018; Khan *et al.*, 2020; Taraky *et al.*, 2021).

There are eight major irrigation systems in the upstream and downstream parts of the Kabul basin (Table 7.2). In the upstream parts of the basin, agricultural lands are often along the river channel. In the upstream part of the Kabul basin, agricultural lands are irrigated using water from canals, streams, springs, karezes and groundwater wells (Rout, 2008; Akhtar *et al.*, 2018).[2] Modern irrigation systems cover 10% of the total irrigated area; karezes and springs account for 5% each, and traditional irrigation systems with intakes from different rivers and riverways account for over 80% (Anderson, 1993; Hashimi *et al.*, 2020). Despite investments in modern irrigation systems in the past two decades, a traditional form of irrigation in which water is diverted from rivers into hand-dug gravity canals remains dominant. Such traditional systems cover a command area of less than 3000 ha (Nathan Associates and Louis Berger International, 1992). Farmers generally lead the development and maintenance of traditional irrigation systems based on local customs and practices with some interventions from the Afghan government. The overall efficiency (i.e. water depleted divided by water applied) of such irrigation systems, including traditional and modern schemes, averages 25–30% (Rout, 2008).

The Nangarhar irrigation project with a command area of 32,000 ha is the largest scheme in the upstream part of the Kabul river basin. The project draws water from the Darunta Dam on the Kabul river near Jalalabad. There are numerous other planned irrigation projects in Afghanistan, including four dams in the Logar

Table 7.1. Land use and land-cover area in Kabul and its sub-basins.

Basin	Area (Mha)	Per cent of basin area							
		Water body	Snow and glacier	Forest	Riverbed	Built-up	Cropland	Barren	Rangelands
Panjsher	0.93	0.14	0.77	0.04	0.13	0.55	6.07	19.78	72.52
Alingor	0.64	0.17	1.00	4.49	0.09	0.35	4.22	13.62	76.05
Logar	0.99	0.00	0.00	0.01	0.02	1.16	5.39	45.68	47.74
Kunnar	2.60	0.17	17.88	6.67	0.34	0.25	2.55	18.44	53.71
Swat	1.53	0.26	9.67	15.38	0.45	0.26	24.16	1.23	48.61
Kabul sub-basin	2.58	0.24	0.03	2.35	1.72	2.53	23.36	19.18	50.59
Kabul	9.26	0.18	6.81	5.39	0.67	1.01	12.64	18.48	54.81

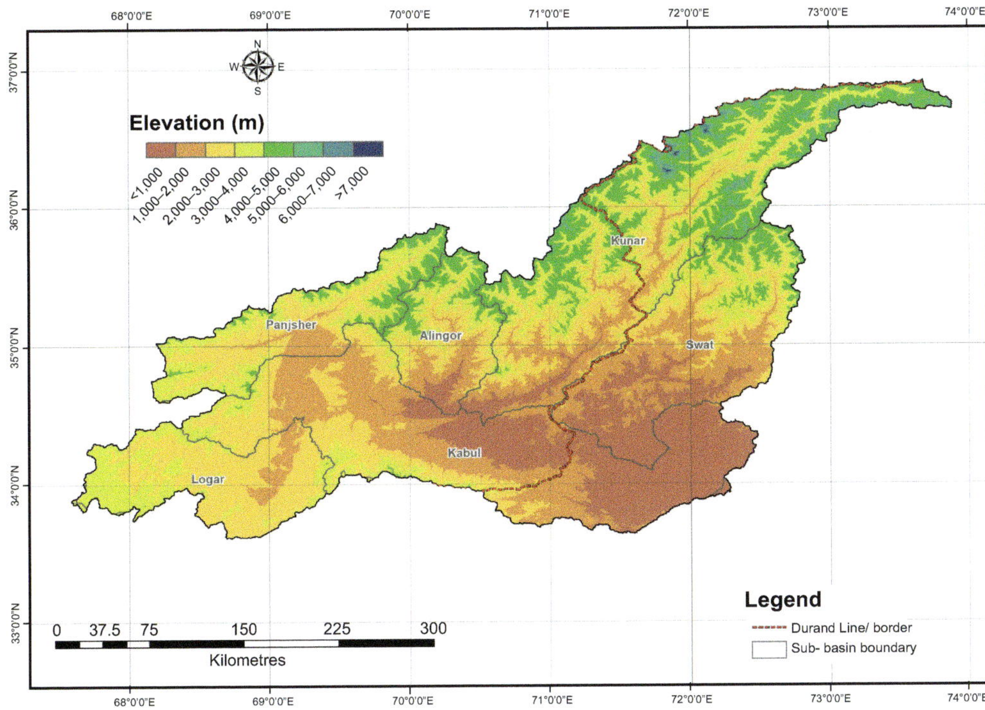

Fig. 7.3. Sub-basins of the Kabul river basin.

Table 7.2. Existing and proposed irrigation systems in the Kabul basin.

Country	Irrigation system	Status	Command area (ha)
Afghanistan	Nangarhar	Existing	32,000
	Logar – upper Kabul	Proposed	105,000
	Panjshir	Proposed	73,000
	Lower Kabul	Proposed	136,000
Pakistan	Upper Swat system	Existing	116,550
	Lower Swat system	Existing	54,430
	Warsak system	Existing	43,470
	Kabul canal system	Existing	29,140

sub-basin, four in Panjshir and five dams and barrages in the lower Kabul sub-basin (World Bank, 2010). These diversion structures could provide surface irrigation to over 300,000 ha of agricultural area. Traditionally, groundwater for irrigation uses an animal-powered wheel to pump water from wells to farms. In recent years, the availability of modern drilling and pumping technologies has led to a significant increase in groundwater use which is most often unregulated, causing large-scale groundwater depletion in some areas (ADB, 2015).

A large percentage of cultivated area in Pakistan is attributed to the existence of multiple large existing irrigation projects (Young *et al.*, 2019). Existing major canal systems within the Kabul river basin in Pakistan (Yang *et al.*, 2013) include: (i) the upper Swat system (diverting from Amandara Headworks on the Swat river); (ii) the lower Swat system (diverting from Munda Headworks on the Swat river); (iii) the Warsak system (diverting water from Kabul river via the Warsak Dam); and (iv) the Kabul canal system (diverting water from the Kabul river). These four systems irrigate around 600,000 ha in Pakistan's portion of the Kabul river basin. Winter wheat and barley (November–May) are the dominant crops in rotation with maize, rice, vines, vegetables and melons in summer (May–November) depending on the weather conditions and water availability (Akhtar *et al.*, 2017). No basin-level crop extents are available. An aggregation of cropped area available at the provincial level from statistical yearbooks for the year 2019–2020 identified around 600,000 ha with wheat and 295,000 ha with rice (PBS, 2019; NSIA, 2021). This is likely an underestimate since the estimates of area under wheat and rice were not available for the newly merged districts (ex-FATA) in Pakistan. In addition, orchard crops such as apples, peaches, almonds, walnuts, apricots and pomegranates are grown perennially and form key export commodities (Fig. 7.4). Despite similar climatic conditions, seven cropping calendars were identified on the Afghanistan side of the Kabul river basin with similarities among administrative units rather than hydrological units. Trends in agricultural area were identified using RCLMS data and verified with other published studies and reveal a meagre increase of 0.5–2% between 2003 and 2018. Non-agricultural land-cover classes also exhibit no decipherable trends. It is plausible that the changes occurred at the level of cropping patterns rather than converting non-agricultural land to croplands.

Kurram river basin

The Kurram river basin covers an area of 2.58 M ha of which 37.5% is in Afghanistan and 62.5% in Pakistan (Fig. 7.5). Rangeland is the dominant land cover (65.14%), with cultivated land (15.33%) being the second most dominant land use. Croplands are concentrated in two clusters. One is in the upstream section of the basin around Khost City in Afghanistan and the second is in the downstream part of the Kurram river basin where the river channel transitions from mountains to plains. This area is served by the Baran Dam located just upstream of Bannu City. Of the total agricultural area in the Kurram, 18.3% is within Afghanistan and 81.7% is within Pakistan. Bare land and riverbeds constitute 8% each. New water storage and diversion projects are planned on the Kurram river on the Pakistan side (e.g. the Kurram Tangi Project), which is expected to further increase the area under cultivation (Malik, 2019). There are two large irrigation schemes in the downstream part of the Kurram basin covering 112,504 ha. The Marwat canal system serves 69,000 ha of command area while the Bannu system serves 43,504 ha.

Gomal river basin

The Gomal river basin covers 39,200 km^2 of which 22% lies in Afghanistan and 78% in Pakistan. Most of the Gomal river basin is comprised of a mountainous, arid landscape. Barren land and rangelands constitute 63% and 31%, respectively, of the total basin area (Fig. 7.6). Meanwhile, both forest cover and agricultural land constitute 4.3% of the basin area (204,495 ha). Only 2% of the total agricultural area is in Afghanistan; the remaining 98% lies on the Pakistan side. The upstream Afghan part of the basin spreads across Gomal, Wolmamay, Wazakhwa and Sarobi districts with sparsely populated rural settlements. The agricultural lands in upstream Afghanistan are mostly along the river channels and close to settlements. The large agricultural lands within the basin are concentrated in a few land clusters on the Pakistan side of the Gomal river basin. The largest agricultural cluster in the basin is located before the Gomal confluence with the Indus river in the district of Dera Ismail Khan, Pakistan. Another agricultural cluster is in the Zhob river basin, a major tributary of the Gomal, which flows along the southern catchment boundary, falls in Qilla Saifullah district and traverses north to join the Gomal river near Khajuri Kach in Zhob district. Major urban settlements include Zhob, Qilla Saifullah and Wana, all in Pakistan.

Fig. 7.4. Orchards in Shamozai, Kabul river basin, Pakistan. (From: Kashif Hussain, IWMI-Pakistan)

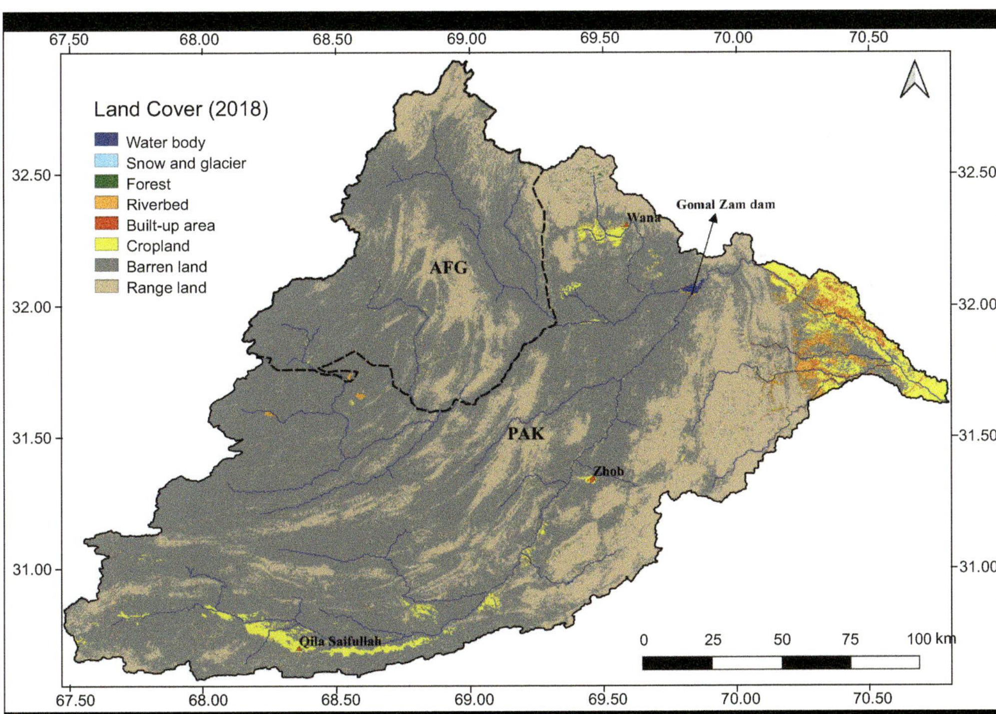

Fig. 7.5. Kurram river basin land cover, 2018.

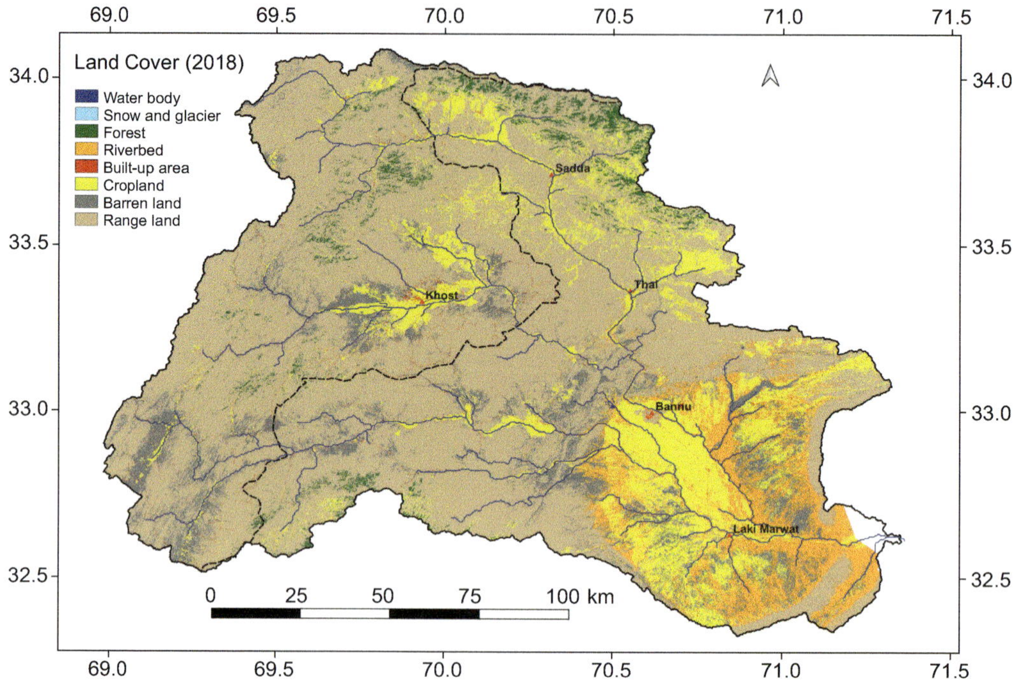

Fig. 7.6. Gomal river basin land cover, 2018.

There is no grey-water storage infrastructure in the Afghan part of the Gomal river basin. The Gomal Zam Dam, located after the confluence of Zhob and Gomal rivers in South Waziristan district (Pakistan), is the largest reservoir in the basin with a capacity of 1410 mcm. The Gomal Zam Command Area Development Project (GZCADP) was envisaged to irrigate 77,295 ha of farmland annually in the districts of Tank and Dera Ismail Khan in Khyber Pakhtunkhwa province during its conceptualization in 2001 (USAID, 2019; Bhatti *et al.*, 2021). The key infrastructure components of the GZCADP include a diversion barrage, 60.5 km main canal and 206 km of distribution canals to facilitate water delivery to the command area. The agricultural area receives its water supply from the Gomal Zam Dam, which enables the cultivation of irrigated wheat and barley crops (Khan *et al.*, 2018). The full potential of the GZCADP is yet to be realized due to delays in infrastructure development. An impact assessment of the existing command area showed increasing yields, production and income as well as on- and off-farm employment opportunities (USAID, 2019).

Agricultural Water Use

Kabul river basin

The long-term average annual rainfall received in the Kabul river basin is about 503.9 mm (Table 7.3). There is large spatial variability in annual average rainfall, which ranges from 140 mm to in excess of 1000 mm. Major parts of the Kabul basin receive rainfall of less than 200 mm per year. However, the southern (Kabul sub-basin) and north-western regions (Swat sub-basin) of the basin in Pakistan form small regions with significantly higher rainfall than most of the basin. With limited rainfall, the prospect of agriculture depends mainly on irrigated water supply for the production of commercial cereal crops. Average water requirements of major crops in Pakistan range from 480 mm for wheat to 1800 mm for sugarcane per annum

Table 7.3. Mean annual precipitation and actual evapotranspiration in Kabul, Kurram and Gomal river basins (2001–2021).

Basin	Afghanistan			Pakistan			Total		
	Precip. (mm)	ETa (mm)	Diff. (mm)	Precip. (mm)	ETa (mm)	Diff. (mm)	Precip. (mm)	ETa (mm)	Diff. (mm)
Kabul	433.9	364.0	69.9	596.1	470.8	125.4	503.9	410.1	93.8
Kurram	409.1	366.2	42.9	514.1	465.7	48.3	472.1	425.9	46.1
Gomal	261.3	225.5	35.8	288.9	261.4	27.5	283.4	254.2	29.2

(Kahlown *et al.*, 2003). The average annual ETa covering all the land-cover classes is around 410 mm.[3] In terms of cropland extent, both Kabul and Swat sub-basins contain approximately 24% of the area under agriculture while the corresponding extent for the other four sub-basins varies from 2% to 6%. There are large spatial variations in the annual average ETa of croplands within the basin (Fig. 7.7a). In the Afghan part of the basin where croplands are concentrated in valleys, the annual average ETa is generally less than 800 mm. Historical estimates of potential ETa in the Afghan part of the basin were approximately 1300 mm per year (Qureshi, 2002). Agricultural areas in the downstream part of the basin in Pakistan have annual average ETa values of over 1000 mm.

While in a similar climate setting, agricultural areas of Afghanistan are at much higher elevations (1700 m masl excluding the Panjsher Valley). Elevation is one of the key factors influencing evapotranspiration (Goulden and Bales, 2014) and the crop water requirement and water use vary significantly between upstream (high-elevation) and downstream (low-elevation) portions of the basin. In addition, cropping patterns between upstream and downstream also play a crucial role in large variations in ETa. While wheat is the dominant crop across the basin, the downstream portion of the basin follows a mainstream crop rotation between wheat–maize or wheat–rice in contrast to wheat with fruit orchids in the central part of the basin in Afghanistan (Yang *et al.*, 2016; Akhtar *et al.*, 2017). In addition, documented evidence points to increased use of groundwater for irrigation throughout Pakistan, including in Khyber Pakhtunkhwa province, which constitutes the downstream part of the basin (Young *et al.*, 2019; Akhtar *et al.*, 2021).

The mean ETa of croplands for the summer season varied from 590 mm to 720 mm over the selected four-year period with an average of 627.5 mm (Fig. 7.7b). Mostly, the variability of seasonal average ETa was consistent at around 600 mm with significant deviations only in 2018. For the Afghan part of the basin, the average ETa for the summer season was 333 mm (Akhtar *et al.*, 2018). In contrast to high ETa values observed in the summer, winter season estimates were within a narrow range of 138–167 mm. The four-year seasonal ETa data and average crop water use in the Kabul basin for the summer season amounts to 7350 mcm. The average ETa for the majority of the summer months ranges between 45 mm and 60 mm while it is less than 20 mm for most of the winter months (Fig. 7.7c). The highest ETa was observed in May, while June–September showed similar monthly ETa estimates. Due to different physiographic conditions between irrigated areas in the Afghan and Pakistani parts of the basin, the monthly ETa values in the downstream part will likely be larger than the average ETa values estimated for the whole basin.

Kurram river basin

The Kurram river basin receives a mean annual rainfall of about 472 mm. Areas with high rainfall are in the mountains in the north and south with the central part of the basin receiving low rainfall. An average annual basin ETa of 410 mm was observed in the Kurram basin encompassing all land-cover types. The croplands are high ETa areas, with values greater than 800 mm but varying up to 1300 mm in most of the observed pixels (Fig. 7.8a). These

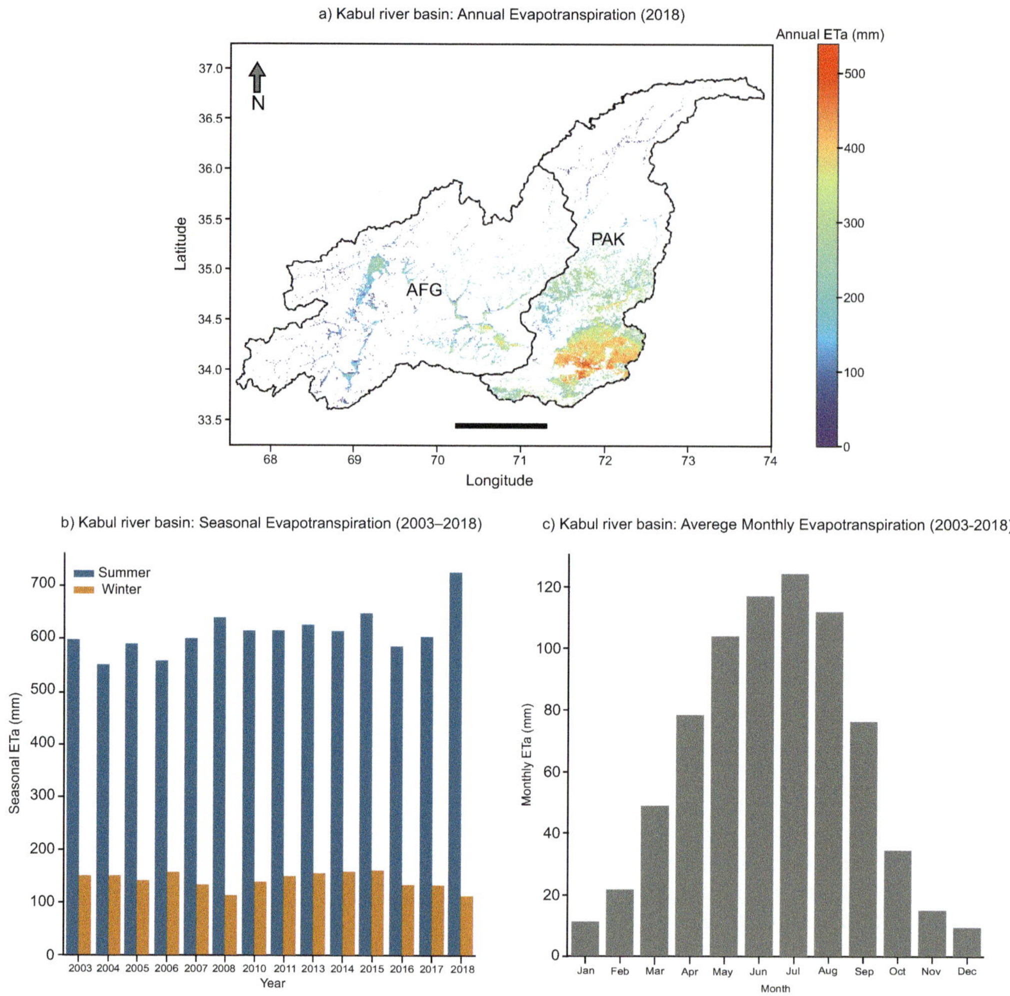

Fig. 7.7. Spatio-temporal variation of actual evapotranspiration (ETa) over croplands in the Kabul river basin: (a) annual average evapotranspiration (2001–2021); (b) seasonal evapotranspiration (2003–2018); (c) average monthly evapotranspiration (2001–2021).

high ETa clusters (mostly in the Pakistan part of the basin) are indicative of intense irrigated agriculture with water from the Baran Dam. In the Afghan part of the basin, a similar high ETa area showing irrigated agriculture is located around the town of Khost and the surrounding river valleys. There are croplands with a comparatively low ETa (<600 mm) on both sides of the basin. These areas may indicate agricultural lands in river valleys or banks, irrigated using traditional systems like mirab. Average seasonal ETa for the summer ranges from 480 mm to 620 mm with 2018 being the year of highest crop water use (Fig. 7.8b). Since the contribution of snow and glacier melt is unknown, it is difficult to estimate the proportion of total available water used for irrigated agriculture. The proportion of crops irrigated with surface and groundwater needs to be ascertained in future basin-level studies. Seasonal average ETa over winter varied from 78 mm to 133 mm. Considering the five-year seasonal ETa data, average crop water use in the Kurram Basin for the summer season amounts to 1890 bcm. Like the other two basins, the monthly average ETa is high in the summer

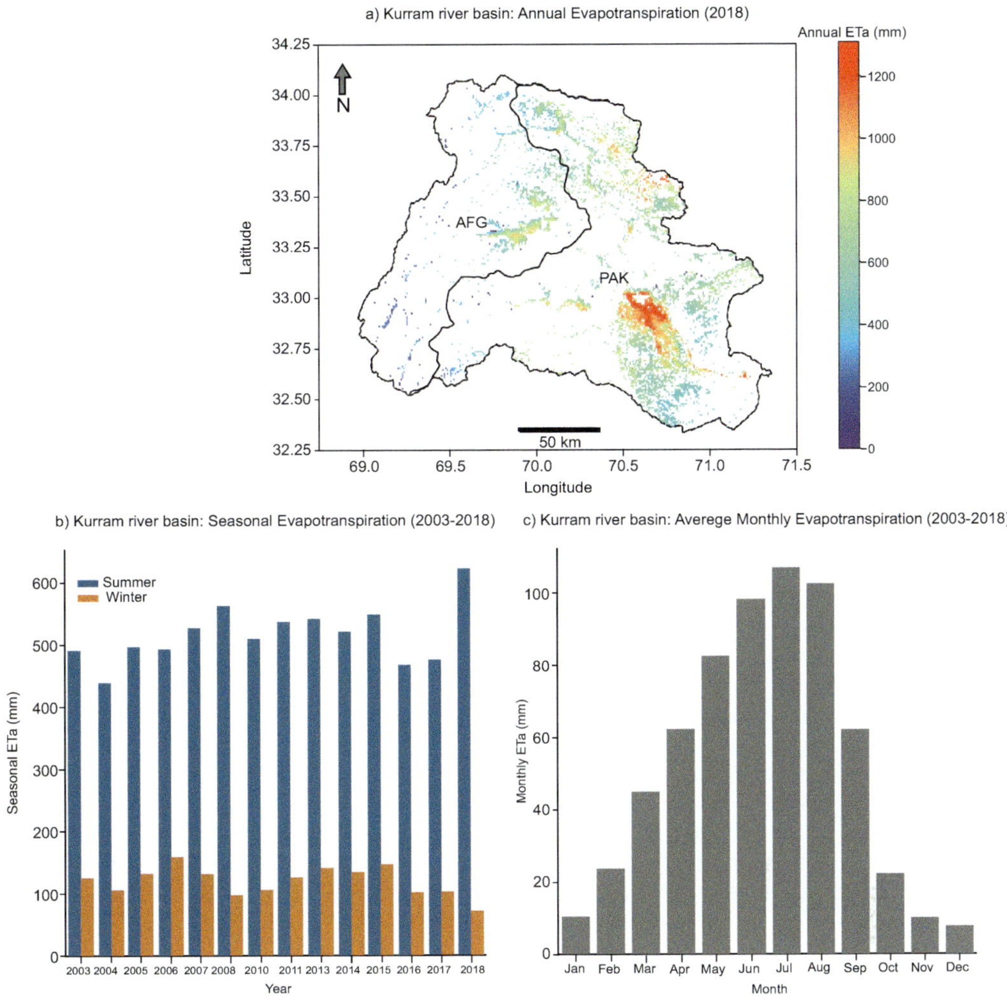

Fig. 7.8. Spatio-temporal variation of actual crop evapotranspiration (ETa) in Kurram river basin croplands: (a) annual average ETa (2001–2021); (b) seasonal ETa (2003–2018); (c) average monthly ETa (2001–2021).

months when snow and glacial meltwater enable irrigated agriculture. The monthly ETa during the summer months varied between 40 mm to 60 mm over croplands in the basin.

Gomal river basin

The Gomal river basin receives an average annual rainfall of about 283 mm (Table 7.3). The basin's annual average ETa is about 225.5 mm. At an aggregate level, there appears to be excess rainfall in the basin. However, a large part of the Gomal basin receives less than 200 mm a year. Higher rainfall values ranging between 350 mm and 400 mm were observed in the south-western and north-eastern mountain regions in the downstream stretch in Pakistan. The available rainfall is not sufficient to meet the water requirements of most commercial crops like wheat or maize. Gomal basin needs irrigation infrastructure to ensure reliable agricultural production in case of intensification plans. Currently, cropland comprises only 4.3% of the basin area, signifying minimal agriculture water use. Most agricultural areas are located in the flat plains of the east where the Gomal river transitions from the mountains and meets the Indus. Most of the Afghanistan part of the Gomal basin comprises uninhabited mountain regions with no agricultural water use. The Gomal Zam Dam serves as the main source of irrigation water in the downstream part of the basin in Pakistan. The average annual ETa for the irrigated areas in the eastern part of the basin varies from 500 mm to 1400 mm (Fig. 7.9a). The areas of high crop water use likely receive irrigation from the Indus river, owing to its proximity, rather than the Gomal Zam Dam.

The average seasonal ETa for the summer months in the croplands was between 360 mm and 580 mm during the period 2014–2018 (Fig. 7.9b). This indicates variability in water availability for irrigation during the summer months in which crop water requirements are usually high. Considering these four years, average agricultural water use in the Gomal river basin accounts for 716 mcm. The average seasonal ETa for the winter season ranges between 80 mm and 100 mm over the croplands and does not show large variability in the summer mainly because of low crop water requirements and a significant portion of cropland remaining fallow. High average monthly ET values were observed from March to August with over 30 mm/month, coinciding with the period of peak crop water requirements (Fig. 7.9c).

Domestic and Industrial Water Use

Kabul river basin

Approximately 34 million people live in the Kabul river basin and consume, on average, 55 lpcd based on the estimates for Kabul City. In rural areas, water consumption may be less. Assuming that 55 lpcd is the average per capita consumption across the basin, annual average domestic water use accounts for 554.6 mcm. The existing domestic water supply system managed by the Afghanistan Urban Water Supply and Sewerage Company serves less than 13.5% of the people through direct connections and another 14% through public water taps. Diversion of water for irrigation severely reduces river flows during summer and Kabul relies mainly on tapping groundwater for domestic use. Currently, the sources for domestic water use are tapped through well fields in the upper Kabul river aquifer, the Paghman and Kharga river aquifer, and the lower Kabul river aquifer. While projects were implemented in the last decade to expand water supply infrastructure, such as the Kabul Urban Water Supply Project, coverage was limited to specific zones.

Kabul City has three industrial parks: Pole Charkhi and Jumma MM Phase 1 and 2. There are also numerous small-scale production facilities in the city. However, there is limited information on water use by these enterprises. As with domestic users, it is likely that they too use groundwater. Most industrial activities in the cities of Peshawar and Mardan are small-scale industries focusing on carpets, garments and textiles. There are no large manufacturing facilities in the Pakistan part of the Kabul basin but as a part of the China Pakistan Economic Corridor, the Rashakai Special Economic Zone (SEZ) is under development near Mardan. This SEZ is expected to host a variety of industries ranging from manufacturing, pharmaceuticals, automobiles and food processing, which will likely draw on surface and groundwater.

Kurram river basin

Like the Gomal, the Kurram river basin consists of predominantly rural settlements. Major urban settlements include Khost in Afghanistan and Bannu in Pakistan. With a population of approximately 5.6 million, Kurram basin users consume approximately 106 mcm per year when per capita consumption of 55 lpcd was used as a per capita estimate. Groundwater is the major component of the domestic water supply in the basin. Piped water supply is provided by the Water and Sanitation Service Company of Bannu within the city limits. Bannu district hosts small industrial estates and leather goods service centres. There are no large manufacturing industries or SEZs in the Kurram basin.

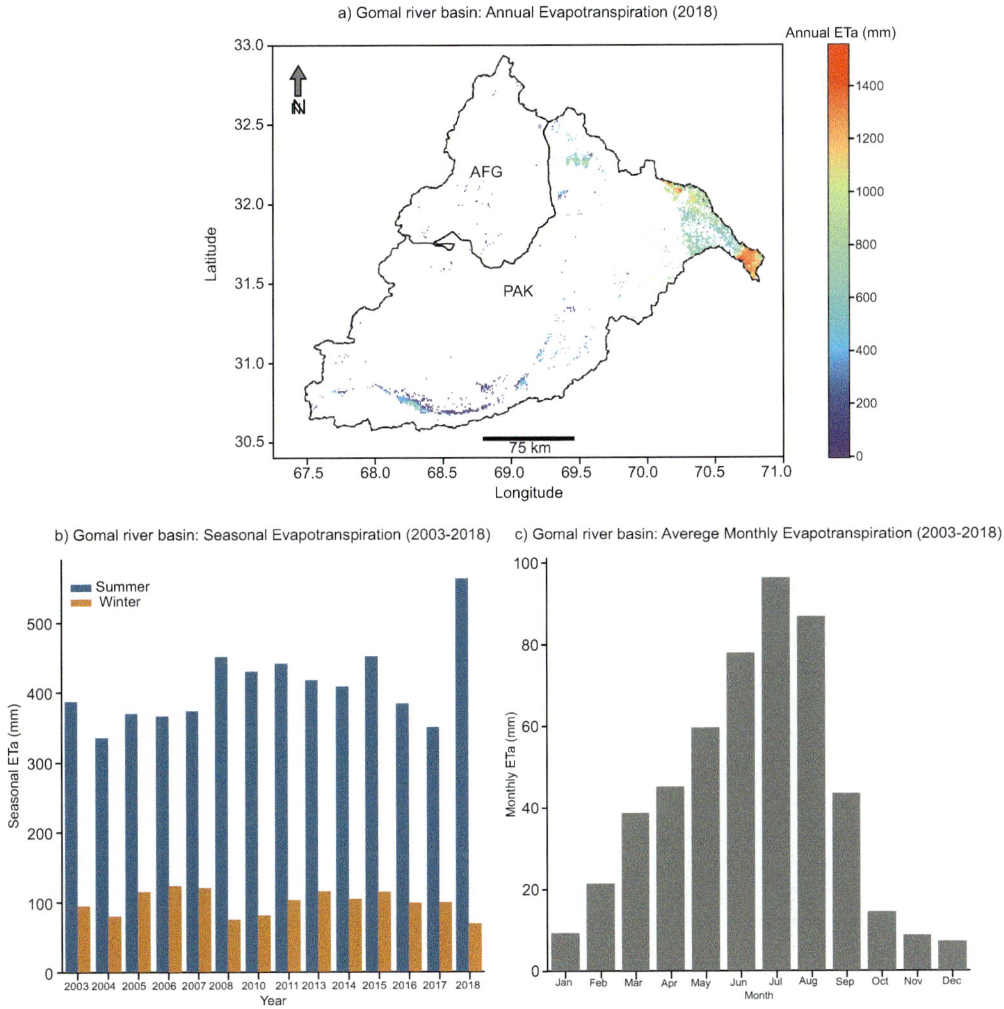

Fig. 7.9. Spatio-temporal variation of actual crop evapotranspiration (ETa) over croplands of the Gomal river basin: (a) annual average evapotranspiration (2001–2021); (b) seasonal evapotranspiration (2003–2018); (c) average monthly evapotranspiration (2001–2021).

Gomal river basin

The population in the Gomal river basin is around 3.8 million of which 98% is on the Pakistan side of the catchment. Most rural and urban settlements are along the Zhob river in the south. Zhob is the largest city and district within the catchment. In Zhob district, protected wells and boreholes form the main source of drinking water (41%) followed by piped water (22%) and unprotected springs (Kakar, 2017). Surface water forms only 6% of the total sources of drinking water supply while most people depend on groundwater. Based on the per capita domestic water use in similar settings in the Afghanistan-Pakistan region, an average value of 55 lpcd was assumed, which resulted in 21 mcm of domestic water use within the catchment. It is expected that per capita water consumption will rise as water delivery services improve. Surface water forms only a minor portion of the current domestic water consumption, with a major contribution from groundwater and traditional springs.

Livestock rearing is a traditional economic activity within the basin, however, there are no large-scale processing units. Most of the livestock sector is dominated by smallholder farming families. No livestock water use estimates have been documented.

The Gomal river basin has no large industrial areas or special economic zones. Even Zhob City, the largest urban settlement within the catchment, remains negligible and underdeveloped (PDD and UNICEF, 2011). Industrial water use is expected to be negligible. However, Bostan Special Economic Zone is currently implemented as a part of the China Pakistan Economic Corridor in Pisthin district bordering Quetta and Balochistan (Aslam *et al.*, 2019). Priority industries include fruit processing and livestock with a focus on Qila Saifullah, which is in the Gomal river basin. These planned industries will likely affect water use indirectly through changes in agricultural production rather than for direct industrial water use.

Challenges and Opportunities for Integrated Basin Management

In the absence of other basin studies on agricultural water use, this chapter leveraged remote-sensing global ETa data sources. The ETa used in this chapter is a coarse resolution dataset having 1 km spatial resolution. While these datasets have been successfully used to undertake studies across global to basin scales, it is recognized that large differences exist among a wide range of ET products due to uncertainties in input data, algorithm limitations and dependency (Wagle and Gowda, 2019). Ideally, these datasets need to be validated against field observations from lysimeter or eddy flux tower data. However, considering the scarcity of ETa datasets from the field in Kabul, Kurram and Gomal river basins, these global datasets were used to identify spatio-temporal patterns, particularly over croplands, and may be used for future basin-level studies estimating agricultural water use.

Annual average water consumption within the three basins relative to observed rainfall and surface water outflows suggests additional benefits can be harnessed. Table 7.4 indicates that 80–86% of the total rainfall goes towards satisfying current basin water consumption, which includes use for domestic, industrial and ETa needs. The annual average runoff measured at basin outlets in these three catchments indicates significant untapped potential, particularly in the Kabul basin (Rehman *et al.*, 1997). These excess flows observed at basin outlets beyond the estimates derived from rainfall minus total consumption indicate a contribution from snow and glacier melt.

Land and water use are skewed towards Pakistan in all three basins. Pakistan uses a higher proportion of water, particularly in the Kabul and Kurram basins. The Kabul basin in particular has a well-established surface water irrigation network in Pakistan that reliably provides water for agricultural consumption and groundwater recharge to a large proportion of farmlands. Moreover, the Kabul river flows contribute significantly to water availability in canals within the Indus Basin Irrigation System (IBIS) outside Kabul basin's extents. The IBIS of Pakistan has two canal subsystems, Jhelum-Chenab Zone and Indus Zone. The latter includes all canal systems within Sindh, south-west Punjab and Balochistan provinces). Jhelum and Chenab rivers are primary water contributors to the Jhelum-Chenab Zone and the Indus and Kabul rivers are primary contributors to the Indus Zone (Siddiqi *et al.*, 2018). The Kabul river contributes approximately 25% of the Indus river's annual discharge at the confluence in Nowshera (Akhtar *et al.*, 2021; Anderson *et al.*, 1993). Thus, Kabul's contribution to water availability for IBIS is critical, especially in the winter months for downstream users during low flow periods (Akhtar and Iqbal, 2017).

The skewed nature of water use in Pakistan poses a critical challenge for both countries. Despite the recent turmoil and uncertainties in the political landscape of Afghanistan, leveraging water availability in the Kabul river basin for agriculture and hydropower to meet food and energy demands will become inevitable. However, in the absence of any bilateral water sharing treaties between Afghanistan and Pakistan and given Pakistan's heavy reliance on the Kabul river flows within and outside the basin, negotiations over water sharing will remain contentious.

Table 7.4. Long-term averaged annual water consumption relative to long-term system outflows in mcm (surface water outflows). (Basin outflow derived from Rehman *et al.*, 1997)

Basin	Rainfall (mcm)	Basin ETa (mcm)	Domestic (mcm)	Industrial (mcm)	Total basin water consumption (mcm)	Rainfall-total consumption (mcm)	Basin outflow (mcm)
Gomal	11,093	8875	21	4	8900	2197	863
Kabul	46,735	39,770	554	115	40,439	6296	23,770
Kurram	12,177	10,250	106	21	10,377	1800	776

In the upstream parts of the Kurram and Gomal basins in Afghanistan, the steep topography is not conducive to large-scale irrigation projects, except in valleys and along river courses. Future infrastructure developments in these parts will likely focus on using river water for hydropower generation and small- to medium-scale expansion of irrigated areas along the valleys where possible. In such cases, a drastic decline in river discharge to downstream Pakistan is not expected due to upstream developments in the Afghan part of Gomal and Kurram basins, and farmers downstream in Pakistan can continue to expect the current share of water. Both basins are less contentious due to the reduced scope for upstream agricultural intensification and this could provide a path towards initiating transboundary co-operation instead of negotiations in the contentious Kabul river basin impeding any progress. However, this needs to be preceded by efforts to collect hydrological data on Gomal and Kurram tributaries bordering these two countries to create a knowledge base for enhancing data sharing, confidence-building measures and a common understanding of water resources availability and use.

Data scarcity remains a key challenge in land- and water use assessment for a shared vision between Afghanistan and Pakistan regarding the management of the Kabul, Kurram and Gomal basins. Despite data limitations, this research confirms that water use in these basins is significantly greater in Pakistan. However, food, energy and water demands in Afghanistan may significantly change the current water use dynamics. Hence, it is imperative that both countries work towards developing a unified framework for land- and water use assessment. Such a framework may lay the foundation for shared and collaborative management of these basins.

Notes

[1] This estimate is similar to the previous estimate of 9% derived from MODIS data (Akhtar *et al.*, 2017).

[2] A kareze is an underground channel through which water flows from an aquifer within mountains to open fields.

[3] This estimate is similar to the previous ETa estimate of 420 mm estimated primarily using MODIS remote-sensing data and surface energy balance model (Akhtar *et al.*, 2018).

Acknowledgements

The authors would like to acknowledge SERVIR HKH for the RLCMS data and Elnashar *et al.* (2021) for making the synthesized global ET product available.

References

ADB (2015) *Preparation of the Afghanistan Water Resource Sector Development Strategy*. Vol. 2 annexes. TA 7994 AFG. Asian Development Bank (ADB), Manila, Philippines.

Akhtar, F., Awan, U.K., Tischbein, B. and Liaqat, U.W. (2017) A phenology based geo-informatics approach to map land use and land cover (2003–2013) by spatial segregation of large heterogenic river basins. *Applied Geography* 88, 48–61. DOI: 10.1016/j.apgeog.2017.09.003.

Akhtar, F., Awan, U., Tischbein, B. and Liaqat, U.W. (2018) Assessment of irrigation performance in large river basins under data scarce environment: a case of Kabul River Basin, Afghanistan. *Remote Sensing* 10(6), 972. DOI: 10.3390/rs10060972.

Akhtar, S.M. and Iqbal, J. (2017) Assessment of emerging hydrological, water quality issues and policy discussion on water sharing of transboundary Kabul River. *Water Policy* 19(4), 650–672. DOI: 10.2166/wp.2017.119.

Akhtar, T., Kahn, H.F. and Mustafa, D. (2021) Water security in Pakistan: availability, accessibility and utilisation. In: Watto, M.A., Mitchell, M. and Bashir, S. (eds) *Water Resources of Pakistan*. Springer International Publishing, pp. 57–78. DOI: 10.1007/978-3-030-65679-9_4.

Anderson, I.M. (1993) *FAO Program for the Rehabilitation of Afghanistan: Rehabilitation of Informal Irrigation Systems in Afghanistan*. Design manual. Available at: http://afghandata.org:8080/xmlui/handle/azu/3553 (accessed 24 July 2022).

Anwar, A.A. and Bhatti, M.T. (2018) Pakistan's water apportionment accord of 1991: 25 years and beyond. *Journal of Water Resources Planning and Management* 144(1), 05017015. DOI: 10.1061/(asce)wr.1943-5452.0000831.

Archer, D.R., Forsythe, N., Fowler, H.J. and Shah, S.M. (2010) Sustainability of water resources management in the Indus Basin under changing climatic and socio-economic conditions. *Hydrology and Earth System Sciences* 14(8), 1669–1680. DOI: 10.5194/hess-14-1669-2010.

Aslam, H., Sakib, M. and Sandhu, A. (2019) SEZs for sustainable development in Pakistan: building on the lessons from China. *Sustainable Development Policy Institute*. Available at: https://think-asia.org/bitstream/handle/11540/11373/SEZsforSustainableDevelopmentinPakistan.pdf?sequence=1 (accessed 24 July 2022).

Bhatti, M.T., Ashraf, M. and Anwar, A.A. (2021) Soil erosion and sediment load management strategies for sustainable irrigation in arid regions. *Sustainability* 13(6), 3547. DOI: 10.3390/su13063547.

Danish, M.S.S., Senjyu, T., Sabory, N.R., Danish, S.M.S., Ludin, G.A. *et al.* (2017) Afghanistan's aspirations for energy independence: water resources and hydropower energy. *Renewable Energy* 113, 1276–1287. DOI: 10.1016/j.renene.2017.06.090.

Elnashar, A., Wang, L., Wu, B., Zhu, W. and Zeng, H. (2021) Synthesis of global actual evapotranspiration from 1982 to 2019. *Earth System Science Data* 13(2), 447–480. DOI: 10.5194/essd-13-447-2021.

Faruqee, T. (1997) Future of irrigation and drainage in Pakistan. *The Pakistan Development Review* 36 (4 Part II), 565–591. DOI: 10.30541/v36i4IIpp.565-591.

Frischmann, P. (2012) *Afghanistan Resource Corridor Development: Water Strategy*. Final Kabul River Basin Report. World Bank. Available at: https://documents1.worldbank.org/curated/en/599831468196445996/text/797210WP0v205-0Box0379789B00PUBLIC0.txt (accessed 15 December 2021).

Goulden, M.L. and Bales, R.C. (2014) Mountain runoff vulnerability to increased evapotranspiration with vegetation expansion. *Proceedings of the National Academy of Sciences* 111, 14071–14075. DOI: 10.1073/pnas.1319316111.

Hashimi, R., Matsuura, E. and Komatsuzaki, M. (2020) Effects of cultivating rice and wheat with and without organic fertilizer application on greenhouse gas emissions and soil quality in Khost, Afghanistan. *Sustainability* 12(16), 6508. DOI: 10.3390/su12166508.

Kahlown, M.K., Ashraf, M., Raoof, A. and ul Haq, X. (2003) Determination of crop water requirements of major crops under shallow water table conditions. Pakistan Council of Research in Water Resources, Islamabad, Pakistan. Available at: https://pcrwr.gov.pk/wp-content/uploads/2020/Water-Management-Reports/Determination%20of%20Crop%20Water%20Requirment%20of%20Major%20Crops%20Under%20Shallow%20Water-Table%20Condition.pdf (accessed 28 July 2022).

Kakar, A. (2017) *Zhob District Profile*. Balochistan Rural Development and Community Empowerment Program 2017–2022. Available at: http://www.rspn.org/wp-content/uploads/2018/04/District-Prof.-_Zhob.pdf (accessed 21 November 2021).

Khan, H.F., Yang, Y.C.E. and Wi, S. (2020) Case study on hydropolitics in afghanistan and pakistan: energy and water impacts of kunar river development. *Journal of Water Resources Planning and Management* 146(9), 05020015. DOI: 10.1061/(asce)wr.1943-5452.0001262.

Khan, S.A., Shah, S.A., Ullah, I., Ibrahim, M. and Khan, S. (2018) Impact of gomal zam dam irrigation project on agriculture and welfare of farming community in southern districts of khyber pakhtunkhwa-pakistan. *Asian Journal of Agriculture and Rural Development* 7(10), 212–218. DOI: 10.18488/journal.1005/2017.7.10/1005.10.212.218.

Latif, Y., Ma, Y. and Ma, W. (2021) Climatic trends variability and concerning flow regime of Upper Indus Basin, Jehlum, and Kabul river basins, Pakistan. *Theoretical and Applied Climatology* 144(1–2), 447–468. DOI: 10.1007/s00704-021-03529-9.

Malik, T. (2019) Pak-Afghan water issue: a case for benefit-sharing. *Policy Perspectives* 16(1), 77–98. DOI: 10.13169/polipers.16.1.0077.

Mu, Q., Zhao, M. and Running, S.W. (2011) Improvements to a MODIS global terrestrial evapotranspiration algorithm. *Remote Sensing of Environment* 115, 1781–1800. DOI: 10.1016/j.rse.2011.02.019.

Najmuddin, O., Deng, X. and Siqi, J. (2017) Scenario analysis of land use change in Kabul River Basin – a river basin with rapid socio-economic changes in Afghanistan. *Physics and Chemistry of the Earth, Parts A/B/C* 101, 121–136. DOI: 10.1016/j.pce.2017.06.002.

Nathan Associates and Louis Berger International (1992) Afghanistan water constraints overview analysis, Nathan Berger Afghanistan studies project. Available at: https://pdf.usaid.gov/pdf_docs/PNABL957.pdf?simple=True (accessed 19 May 2022).

NSIA (2021) *Afghanistan Statistical Yearbook 2020*. National Statistics and Information Authority. Available at: https://invest.gov.af/theme3/wp-content/uploads/2021/06/Afghanistan-Statistical-Yearbook-first-Version.pdf (accessed 24 July 2022).

PBS (2019) *Pakistan Statistical Yearbook 2019*. Pakistan Bureau of Statistics, Islamabad, Pakistan. Available at: https://www.pbs.gov.pk/sites/default/files/other/Pakistan_Statistical_yearbook_2019.pdf (accessed 24 July 2022).

PDD and UNICEF (2011) *Zhob District Development Profile*. Available at: https://cms.ndma.gov.pk/storage/app/public/publications/January2021/05HfnykCroOSZ2pKXZGU.pdf (accessed 24 July 2022).

Qureshi, A.S. (2002) Water resources management in Afghanistan: the issues and options. *IWMI Working Paper 49*. International Water Management Institute (IWMI), Colombo, Sri Lanka. Available at: https://www.iwmi.cgiar.org/publications/iwmi-working-papers/iwmi-working-paper-49/ (accessed 8 May 2022).

Qureshi, A.S., Akhtar, M. and Shah, T. (2004) Role of changing energy pricing policies on groundwater development in Pakistan. *Journal of Applied Irrigation Science* 39(2), 329–342.

Qureshi, A.S., Gill, M.A. and Sarwar, A. (2010) Sustainable groundwater management in Pakistan: challenges and opportunities. *Irrigation and Drainage* 59, 107–116. DOI: 10.1002/ird.455.

Qureshi, R.H. and Ashraf, M. (2019) *Water Security Issues of Agriculture in Pakistan*. Pakistan Academy of Sciences, Islamabad. Available at: https://www.paspk.org/wp-content/uploads/2019/06/PAS-Water-Security-Issues.pdf (accessed 24 July 2022).

Rehman, S.S., Sabir, M.A. and Khan, J. (1997) Discharge characteristics and sediment load from rivers of northern indus basin. *Geological Bulletin (University of Peshawar)* 30, 325–336. Available at: http://nceg.uop.edu.pk/GeologicalBulletin/Vol-30-1997/Vol-30-1997-Paper25.pdf

Rout, B. (2008) How the water flows: a typology of irrigation systems in Afghanistan. Afghanistan Research and Evaluation Unit issue paper series. Available at: https://cropwatch.unl.edu/documents/WOL_Irrigation_Typology_Afghanistan.pdf (accessed 24 September 2022).

Saah, D., Tenneson, K., Matin, M., Uddin, K., Cutter, P. *et al.* (2019) Land cover mapping in data scarce environments: challenges and opportunities. *Frontiers in Environmental Science* 7. DOI: 10.3389/fenvs.2019.00150.

Saleem, S. and Raouf, M.E. (2011) Sustainable agricultural development and the challenges facing agricultural education in Afghanistan. *Journal of Developments in Sustainable Agriculture* (6), 45–49. DOI: 10.11178/jdsa.6.45.

Sharma, B., Sikka, A.K., Sah, R.P. and Cai, X. (2016) Agriculture and water use: implications for sustainable intensification. In: Bharathi, L., Sharma, B. and Smakhtin, V. (eds) *The Ganges River Basin Status and Challenges in Water, Environment and Livelihoods*. Routledge & CRC Press, pp. 93–113.

Sibtain, M., Li, X., Bashir, H. and Azam, M.I. (2021) Hydropower exploitation for pakistan's sustainable development: a SWOT analysis considering current situation, challenges, and prospects. *Energy Strategy Reviews* 38, 100728. DOI: 10.1016/j.esr.2021.100728.

Siddiqi, A., Wescoat, J.L. and Muhammad, A. (2018) Socio-hydrological assessment of water security in canal irrigation systems: a conjoint quantitative analysis of equity and reliability. *Water Security* 4–5, 44–55. DOI: 10.1016/j.wasec.2018.11.001.

Syed, A., Raza, T., Bhatti, T.T. and Eash, N.S. (2022) Climate Impacts on the agricultural sector of Pakistan: Risks and solutions. *Environmental Challenges* 6, 100433. DOI: 10.1016/j.envc.2021.100433.

Taraky, Y.M., McBean, E., Liu, Y., Daggupati, P., Shrestha, N.K. *et al.* (2021) The role of large dams in a transboundary drought management co-operation framework: case study of the Kabul River Basin. *Water* 13(19), 2628. DOI: 10.3390/w13192628.

TDAP (2016) Statistics for 2015-2016. Islamabad: Trade Development Authority of Pakistan (TDAP). Available at: http://www.tdap.gov.pk/tdap-statistics.php (accessed 8 May 2022).

Uddin, K., Matin, M.A., Khanal, N., Maharjan, S., Bajracharya, B. *et al.* (2021) Regional land cover monitoring system for Hindu Kush Himalaya. In: Bajracharya, B., Thapa, R.B. and Matin, M.A. (eds) *Earth Observation Science and Applications for Risk Reduction and Enhanced Resilience in Hindu Kush Himalaya Region: A Decade of Experience from SERVIR*. Springer International Publishing, Cham, Switzerland, pp. 103–125. Available at: https://doi.org/10.1007/978-3-030-73569-2_6

UNAMA (2016) *Water Rights: An Assessment of Afghanistan's Legal Framework Governing Water for Agriculture*. United Nations Assistance Mission in Afghanistan. Available at: https://unama.unmissions.org/sites/default/files/2016_19_10_water_rights_final_v2.pdf (accessed 24 July 2022).

USAID (2014) *Planning and Engineering Services for Master Plan in Peshawar Khyber Pakhtunkhwa: Drinking Water, Sanitation/Storm Water and Solid Waste Services*. Executive Summary – Consolidation of Key Findings of Volumes 1–3. Prepared by National Development Consultants and MWH Global Inc. Available at: https://urbanpolicyunit.gkp.pk/wp-content/uploads/2018/02/Water-Sanitation-Peshawar-Report.pdf (accessed 24 July 2022).

USAID (2019) *Gomal Zam Command Area Development Project: Midterm Performance Evaluation*. Available at: https://pdf.usaid.gov/pdf_docs/PA00W7PX.pdf (accessed 24 July 2022).

von Grebmer, K., Bernstein, J., Wiemers, M., Schiffer, T., Hanano, A. *et al.* (2021) Global hunger index: hunger and food systems in conflict settings, welt hunger life and concern worldwide. Available at: https://www.globalhungerindex.org/pdf/en/2021.pdf (accessed 8 May 2022).

Wagle, P. and Gowda, P.H. (2019) Remote sensing of evapotranspiration (ET) editorial. *Remote Sensing* 11, 2146. DOI: 10.3390/rs11182146.

Waqas, M.M., Shah, S.H.H., Awan, U.K., Waseem, M., Ahmad, I. *et al.* (2020) Evaluating the impact of climate change on water productivity of maize in the semi-arid environment of Punjab, Pakistan. *Sustainability* 12, 3905. DOI: 10.3390/su12093905.

Weatherhead, E.K. and Howden, N.J.K. (2009) The relationship between land use and surface water resources in the UK. *Land Use Policy, Land Use Futures* 26, S243–S250.

Wescoat, J.L., Ahmed, W., Burian, S., Punthakey, J.F. and Shahid, A. (2021) Examining Irrigated Agriculture in Pakistan with a Water-Energy-Food Nexus Approach. In: Watto, M.A., Mitchell, M. and Bashir, S. (eds) *Water Resources of Pakistan: Issues and Impacts, World Water Resources*. Springer International Publishing, Switzerland. Available at: https://doi.org/10.1007/978-3-030-65679-9_9

WHO and UNICE (2021) *Progress on Household Drinking Water, Sanitation and Hygiene 2000–2020: Five Years into the SDGs*. World Health Organization and the United Nations Children's Fund, Geneva, Switzerland. Available at: https://www.who.int/publications/i/item/9789240030848 (accessed 24 July 2022).

Williams, M.W. (2013) The status of glaciers in the hindu kush–himalayan region. Available at: http:// books. icimod. org. hardcopies available from: distri@ icimod. org. ISBN978-92-9115-215-5, *Mountain Research and Development* 33(1), 114–115. DOI: 10.1659/mrd.mm113.

World Bank (2010) *Afghanistan-Scoping Strategic Options for Development of the Kabul River Basin: A Multisectoral Decision Support System Approach*. World Bank, Washington, DC. Available at: https://openknowledge.worldbank.org/handle/10986/18422 (accessed 24 July 2022).

WTO (2020) Afghanistan trade profile. Available at: https://www.wto.org/english/res_e/statis_e/daily_update_e/trade_profiles/AF_e.pdf (accessed 8 May 2022).

Yang, Y.-C.E., Brown, C.M., Yu, W.H. and Savitsky, A. (2013) An introduction to the IBMR, a hydro-economic model for climate change impact assessment in Pakistan's Indus River basin. *Water International* 38, 632–650. DOI: 10.1080/02508060.2013.830691.

Yang, Y.C.E., Ringler, C., Brown, C. and Mondal, Md.A.H. (2016) Modeling the agricultural water–energy–food nexus in the Indus river basin, Pakistan. *Journal of Water Resources Planning and Management* 142(12), 04016062. DOI: 10.1061/(asce)wr.1943-5452.0000710.

Young, W.J., Anwar, A., Bhatti, T., Borgomeo, E., Davies, S. *et al*. (2019) *Pakistan: Getting More from Water*. World Bank, Washington, DC. Available at: https://openknowledge.worldbank.org/handle/10986/31160 (accessed 24 July 2022).

8 Institutions and Governance

Muhammad Azeem Ali Shah and Fazilda Nabeel

Chapter Overview

Fostering co-ordinated management of a transboundary basin can benefit from an understanding of the internal, often multilevel governance mechanisms in each country sharing it. This chapter delves into the legal and institutional architecture of water governance in basins shared between Afghanistan and Pakistan. Drawing on a careful document analysis as well as expert interviews, the chapter analyses the national laws and institutions in Afghanistan, as well as the legal and institutional provisions for water governance at the federal, provincial and local levels in Pakistan. A review of the existing design of internal water governance in both Afghanistan and Pakistan reveals that despite the existence of well-meaning and well-structured legal provisions, implementation is a key challenge for effective water governance to be addressed on both sides of the Durand Line. The chapter concludes that strengthening internal institutions for better implementation and engaging multilevel stakeholders for improved co-ordination of water policies between riparian countries can pave the path towards successful transboundary basin management in the Kabul, Kurram and Gomal river basins.

Fig. 8.1. Gathering of informal institution in the Gomal river basin.

DOI: 10.1079/9781800622371.0008

Introduction

Transboundary water governance is influenced by decisions 'within, below, and beyond the interstate levels' (Zeitoun *et al.*, 2016). This matter highlights the significance of water governance mechanisms in place at the national and local levels. Co-operation on transboundary water governance necessarily requires an appreciation of the multiscalar nature of water governance in countries that are home to shared river basins (Moss and Newig, 2010). Multilevel, federal and state institutional arrangements contribute critically and add more complexity to shared basin management (Newig and Fritsch, 2009; Schulz *et al.*, 2017).

In line with this important strand of work, this chapter aims to analyse the national and subnational governance for water in Afghanistan and Pakistan. Both Afghanistan and Pakistan are at a crossroads with regard to their internal water governance. The national, legal and institutional mechanisms for water governance within both countries have a bearing on how opportunities for transboundary governance in the shared Kabul, Kurram and Gomal river basins shape up. The analysis largely focuses on the legal and institutional provisions for water governance as promulgated by the laws of each country; however local, informal ways of governing water have been discussed wherever possible.

Extant studies show how greater alignment of water governance between riparian countries in a shared basin leads to improved co-operation and co-ordination of water resource management (Rawlins, 2019). National water-resource and related institutions, and the formal and informal mechanisms for water governance in basin countries play a key role in preparing them for sustainable transboundary water governance and can be key drivers of co-operation at the basin level (World Bank, 2017).

A number of studies have discussed the need for transboundary co-operation between Afghanistan and Pakistan, pointing to the political economy issues that have stymied shared basin management and benefit sharing in the region (Sadeqinazhad *et al.*, 2018; Shams and Muhammad, 2023). Studies have also analysed the formal and informal mechanisms for national water governance in Afghanistan and Pakistan separately (Young *et al.*, 2019; Yasin *et al.*, 2020). No study to date has conducted a detailed comparative analysis of national legal and institutional structures of water governance in Afghanistan and Pakistan with a goal to highlight the extent to which national water governance can support co-operative basin management in the Kabul, Kurram and Gomal river basins. This chapter aims to fill this important gap and argues for better co-ordination of internal water governance in both countries as an important first step to shared basin management.

An appreciation of the multiscalar nature of water governance in transboundary basins can also shed light on roadblocks to effective co-operation between basin countries as suggested by Earle and Neal (2017). The analysis of national and subnational water governance in each country also points to the need for engaging stakeholders at multiple levels for successful transboundary water governance in the three basins.

Since Afghanistan is a unitary state, analysis of the legal architecture of water governance involved an analysis of national legislation pertaining to water and water rights.[1] Pakistan is a federation with the provinces as federating units; hence this chapter summarizes the important federal laws relating to water and water rights and then summarizes the relevant laws of the province of Khyber Pakhtunkhwa (KPK), the province through which these western transboundary rivers flow and meet the Indus.

The chapter reviews both formal and informal institutions for water governance in Afghanistan and the recent developments in the political governance of the country to discuss implications for the future of water governance on the Afghan side of the western transboundary river basins. For Pakistan, both federal and provincial legal architecture for water governance is presented, with a focus on the legal and institutional set-up of water resource governance in the province of KPK. While there is no formal water sharing agreement between Afghanistan and Pakistan, the chapter briefly reviews the existing governance mechanisms on both sides of the Durand Line that could be used for transboundary co-operation.

The methodology of this chapter is based upon a textual review of the legal, institutional, regulatory and policy framework related to water and environmental governance. In addition to national legal and institutional provisions, international legal principles applicable to transboundary water governance in shared basins were reviewed. The extensive document review was supplemented by semi-structured expert interviews.

Water Laws and Institutions in Afghanistan

Water governance in Afghanistan consists of a complex mix of formal and informal laws and institutions. Formally, the Afghanistan Water Law (2009) and the Water Affairs Management Law (2020) adopted principles of IWRM and established a river basin management system by enrolling several ministries to establish a permit system to regulate water use. Notwithstanding the detailed legal and institutional set-up under the Afghanistan Water Law, in practice most farmers defer to the informal tribal and customary laws, which are administered through local village water masters called *mirabs*.

The Constitution of the Islamic Republic of Afghanistan (2004)

The subject of water was not explicitly mentioned in the 2004 Constitution and the responsibility of the state with regard to water governance was subsumed within its responsibility for the protection of natural resources and the environment. Article 9 of the Constitution said: 'the protection, management and proper use of public properties as well as natural resources shall be regulated by law'. Article 15 provided that '[t]he state shall be obligated to adopt necessary measures to protect and improve forests as well as the living environment'. It is not yet clear what will replace the 2004 Constitution or if water will be addressed.

Afghanistan's Civil Code (1977)

The Civil Code (1977) addresses issues related to water rights and entitlements, declaring that '[w]ater of rivers and their tributaries are public property'.[2] However, '[e]veryone ... has the right to irrigate ... or draw on a stream for irrigation purposes' including 'for irrigation of crops and trees', so long as the use is not 'contrary to public interests or special laws'.[3] The Civil Code does not clarify what types of uses are or are not contrary to public interests, except to note that the 'usage of water from public streams and its distribution shall be exercised with due observation of prevention of harm to public interests and proportionate to the lands that are intended to be irrigated'.[4]

Provisions of the Civil Code are generally consistent with Article 40 of the 2004 Constitution, which protects private ownership of property. These include: 'No one may build a watercourse, without a right to do so'.[5] A '[p]erson who builds [a] special stream in his own property ... [has] the right to use it in any way he wishes and other persons may not use it without the builder'.[6] Where an irrigation canal is lawfully built on another person's property pursuant to a right of way or passage, the other property owner may not deprive the builder of its use.[7] However, a landowner seeking to irrigate his land cannot 'compel owners of lands under water to let the water pass through their lands, unless he has a right of way or passage of water through those lands'.[8] If a right of way or passage for water already exists, the right 'shall remain so...[but] shall be terminated if there is obvious harm'.[9]

Afghanistan's Environment Law (2007)

The Afghanistan Environment Law (2007) addresses the management of issues relating to the rehabilitation of the environment and protection of natural resources, living organisms and non-living organisms.[10] It defines the environment as meaning 'natural resources, interactions between the components of natural resources and between those components and humans and animals, and physical, aesthetic and cultural qualities that may affect the health and well-being of humans'.[11]

Some key provisions of the law relate to environmental considerations relevant to water resource governance. According to Article 34, all persons dealing with water resources are to take into account the protection of aquatic resources, associated ecosystems, biological diversity and the reduction and prevention of pollution and degradation of water resources.[12] All ministries and national institutions preparing water resource management plans must take into account (i) the provisions for integrated watershed management; (ii) regulation of sustainable abstraction of groundwater; (iii) regulation of the use of surface water for agricultural, industrial, mining and urban purposes; (iv) measures to protect human health and ecosystems; (v) measures to protect wetlands and their associated ecosystems; and (vi) any other provisions necessary for the sustainable use and management of water resources.[13] Any owner or occupier of land on which activity likely to cause significant environmental pollution of a water resource is planned must take certain measures to prevent such pollution, including complying with prescribed waste standards or pollution management practices or ceasing the activity, eliminating the source of pollution and remedying any effect or disturbance to the banks of a watercourse.[14]

Afghanistan's Water Law (2009)

Afghanistan's Water Law (2009) is based on the principles of IWRM and uses a river basin approach to water resource governance, whilst also integrating informal institutions to support water governance at the local level. This section reviews the legal provisions and institutional architecture for water provision under the 2009 law. This law necessitated the adoption of the IWRM approach and a reform of existing water institutions and the establishment of river basin agencies, river basin councils, water user associations and irrigation associations. A mapping of the institutions responsible for water governance under the Water Law (2009) is given in Table 8.1 and Fig. 8.2.

The Afghanistan Water Law (2009) (the Act) was published in the official Gazette No. 980 in 2009 and enacted 'for the purpose of conservation, equitable distribution, and the efficient and sustainable use of water resources'.[15] The Act declares, 'Water belongs to the public and government is responsible for its protection and management'.[16] It supersedes the previous law for water resource use published in the official Gazette No. 755 in 1370.[17]

The Act states that integrated water resource management and development is to be carried out using a river basin approach.[18] The development, storage, use, control, and conservation of water resources are to be guided by the National Water Policy and strategy and environmental protection.[19] The uses of water are prioritized 'with due consideration for the praiseworthy customs and traditions of the people to meet the needs of drinking water, livelihoods, agriculture, industry, public services, energy production, transportation, navigation, fisheries and the environment' with priority for use of water resources given to drinking water and livelihoods'.[20] The Act declares, 'water use is free' but that water service providers may charge users a fee for the supply, storage, transmission, diversion, treatment, operation, and maintenance of the water supply and irrigation systems.[21] Under the Water Law (2009), the institutional responsibilities for water governance are defined as set out in Table 8.1.[22]

Chapter 2 of the Act relates to the management of the use of water resources and stipulates additional duties and responsibilities of the Ministry of Energy and Water (MEW) and Ministry of Agriculture, Irrigation and Livestock (MAIL).[23]

Article 9 of the Act establishes a Supreme Council of Water Affairs Management (SCWM) responsible for co-ordinating and revising facilities in the implementation of water affairs. Chapter 3 of the Act establishes river basin agencies, river basin councils, river sub-basin councils and water users associations 'formed with the purpose of integrated water resources planning and participation of the users and other social and cultural institutions in the decision-making process for management and development of water resources, protection of the environment, equitable distribution of water, and other water matters'.[24] River basin

Table 8.1. Responsibilities of Afghan institutions under Afghanistan's Water Law (2009).

Responsibility	Relevant government institutions
Protection, control, management and effective use of the water of the country	Government of Afghanistan
Planning, management and development of water resources	Ministry of Energy and Water
Planning and implementing activities to survey, investigate, research and assess groundwater reserves and their protection from pollution	Ministry of Mines with co-operation from (i) Ministry of Public Health; (ii) National Environmental Protection Agency
Protection and control of surface water from pollution and monitoring its quality	National Environmental Protection Agency with co-operation from: (i) Ministry of Agriculture, Irrigation and Livestock; (ii) Ministry of Energy and Water; (iii) Ministry of Urban Development; (iv) Ministry of Rural Rehabilitation Development; (v) Ministry of Public Health; (vi) Ministry of Mines
Determination of irrigation norms in different river basins, irrigation drainage systems and other related research for water use for agriculture and irrigation	Ministry of Agriculture, Irrigation and Livestock with co-operation from: (i) Ministry of Energy and Water; (ii) Ministry of Transport and Aviation; (iii) Ministry of Public Health; (iv) National Environmental Protection Agency
Provision of water supply for drinking and livelihoods, including construction of water treatment plants, water conveyance facilities, sewerage systems and sewage treatment plants	Ministry of Urban Development with co-operation from: (i) Ministry of Energy and Water; (ii) Ministry of Mines; (iii) Ministry of Public Health; (iv) Ministry of Agriculture, Irrigation and Livestock; (v) National Environmental Protection Agency
Provision of drinking water supplies and sewage treatment systems in villages by government or non-governmental organizations and construction of small water infrastructure for various village uses	Ministry of Rural Rehabilitation and Redevelopment with co-operation from: (i) Ministry of Energy and Water; (ii) Ministry of Public Health; (iii) Ministry of Mines; (iv) Ministry of Agriculture, Irrigation and Livestock; (v) Ministry of Urban Development; (vi) National Environmental Protection Agency
Rights of way for water resources and water infrastructure, storage facilities, diversions, rivers, traditional and engineered canals, karezes, springs, wells and other related natural water courses in light of the principles of Islamic jurisprudence	Ministry of Energy and Water with co-operation from: (i) Ministry of Agriculture, Irrigation and Livestock; (ii) Ministry of Mines; (iii) Ministry of Rural Rehabilitation and Development; (iv) related but unspecified departments
Management and planning for transboundary waters between Afghanistan and its neighbouring countries and changes of watercourses	Ministry of Energy and Water with agreements from: (i) Ministry of Foreign Affairs; (ii) Ministry of Interior; (iii) Ministry of Border and Tribal Affairs
Maintenance of a balanced record of water consumption and wastewater disposal	Institutions providing water supply and wastewater disposal services 'shall be submitted to the pertinent administrative authorities when required'.

agencies have the duty and authority to develop plans and manage water in accordance with the national water resource policies in line with the characteristics and needs of the relevant river basin.[25] River basin councils are to be established by the Ministry of Energy and Water and may be delegated some of the powers of the Ministry. The duties of river basin councils include determining water allocations, in particular for river basins in accordance with the National Water Policy 2018, managing and monitoring water rights, issue, register, modify or cancel permits,

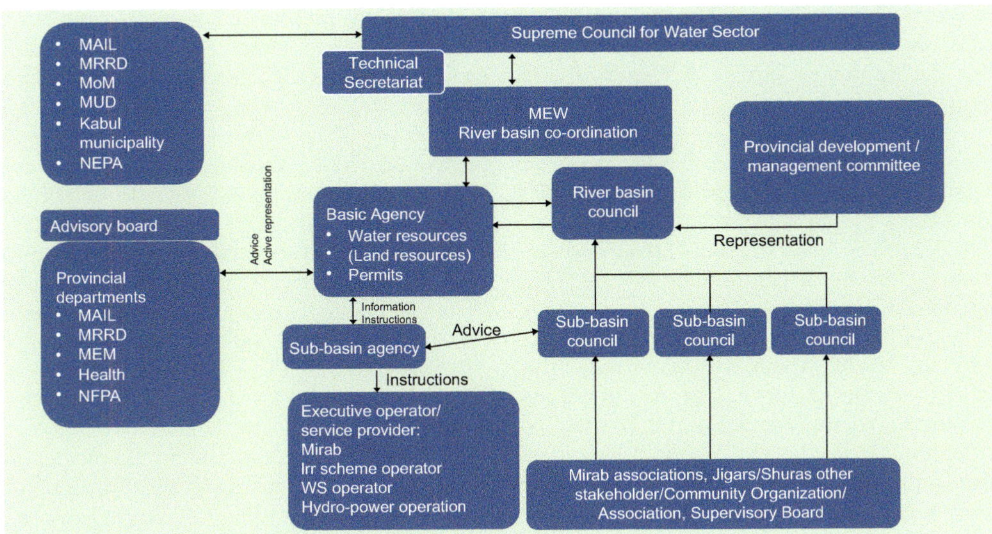

Note: MAIL = Ministry of Agriculture, Irrigation and Livestock, MEW = Ministry of Energy and Water, MoM = Ministry of Mines, MRRD = Ministry of Rural Rehabilitation and Development, MUD = Ministry of Urban Development, NEPA = National Environmental Protection Agency

Fig. 8.2. Institutions responsible for water governance in Afghanistan. (Adapted from Rasooly, 2019)

and resolve disputes regarding the distribution of water.[26] Decisions of river basin councils may be implemented by river basin agencies[27] and may be appealed before the Ministry of Energy and Water and thereafter before the courts.[28] Water user associations can be established under this Act to carry out such duties and functions as prescribed in their respective charters.[29]

Chapter 4 of the Act relates to the use of water and provides for a system of permits regulating its use. Article 19 prohibits the use of water resources without a permit except for (i) drinking water, livelihood and other needs (up to 5 m^3 per household); (ii) navigation; (iii) fire extinguishing; and (iv) limited existing water rights. An activity permit or usage licence from a river basin council is required for surface and groundwater use for newly established development projects, disposal of wastewater and drainage water into water resources, use of water for commercial and industrial purposes, digging and installation of shallow and deep wells for commercial, agricultural, industrial and urban water supply purposes, construction of dams or storage structures if storage capacity exceeds 10,000 m^3, as well as the construction of structures that encroach on the banks, beds, courses or protected rights of way of streams, wetlands, karezes and springs.

The responsibility of regulating the distribution of water within the irrigation networks of designated areas may be delegated by the Ministry of Agriculture, Irrigation and Livestock to irrigation associations who may, in turn, delegate such responsibilities to a Head Water Master (*Mirab Bashi*) or Water Master (*Mirab*) designated as such by the irrigation associations.[30] Water quality standards for drinking water and sanitation, agriculture and industrial wastewater are to be established by the Ministry of Public Health, the Ministry of Agriculture, Irrigation and Livestock, and the Ministry of Urban Development, respectively.[31] Use of water for micro-hydropower generation may be done with agreement from the Ministry of Agriculture, Irrigation and Livestock, the Ministry of Rural Rehabilitation and Development, and irrigation associations.[32] Requests for the use of rivers, canals or other water resources for transportation and navigation are to be submitted by the river basin agency concerned before the Ministry of Energy and Water and then endorsed by the Supreme Council of Water Affairs Management.[33]

According to Chapter 5 of the Act, the National Environment Protection Agency and the Ministry of Public Health are designated to maintain tolerance limits for pollution of water resources. The Ministry of Energy and Water (MEW) is responsible for dams after approval from the Supreme Council of Water Affairs Management.

While the Water Law has established formal institutions for water governance in Afghanistan, it also upholds and works with the informal traditional *mirab* system. According to Chapter 6 of the Act, if a dispute arises amongst farmers within an irrigation network, they are to be resolved by the irrigation association with the help of the Head Water Master (*Mirab Bashi*) and Water Master (*Mirab*) within two days. The Act then provides an appellate process that leads from the river basin council to the Ministry of Agriculture, Irrigation and Livestock and then to the courts.[34] Tampering with, damaging or destroying water resources may be punished with imprisonment of up to two years and a fine commensurate with such tampering, damage or destruction.[35]

The 2009 Water Law also delineates the responsibility for research, identification and protection of groundwater to the MEW in close co-ordination with MoPH and NEPA (Articles 8 and 9). MEW also has the responsibility to license and issue permits for extraction of more than 5 m^3 per day under Articles 19 and 21. Yet in many places, including Kabul, private commercial extraction of groundwater continues without any regulation or permit from MEW.[36] Despite groundwater governance being an integral component of the Water Law (2009), in practice the issue of groundwater is side-stepped and considered in isolation from the river basin agencies.[37]

Afghanistan's Water Affairs Management Law (2020)

A new national statute, the Water Affairs Management Law (2020), replaced the Water Law (2009). The new Water Law aims to promote equitable, efficient and sustainable water use and attempts to harmonize water sector policies and institutions.[38] While the new legislation clarifies responsibilities and affirms the authority of the Supreme Council of Water, Land and Environment (SCWLE), it also established the National Water Affairs Regulatory Authority (NWARA) by presidential decree (Table 8.2). According to the Water Affairs Management Law (2020), NWARA assumed many responsibilities previously vested with the MEW and acted as the office in charge of water resource governance. The Water Affairs Management Law (2020) empowers NWARA to determine sectoral water allocations every three years for drinking, agriculture, industry, environment, energy and tourism. NWARA was also responsible for managing regional data on meteorological and hydrological stations. The 2020 Water Affairs Management Law assigned environmental planning to SCWLE and established a Water Affairs Management Coordination Committee to harmonize sectoral planning and implementation. However, the structure and functions of this committee have yet to be defined. However, due to the political change in August 2021, the Islamic Emirates of Afghanistan (the new regime) has reinstated the MEW with all of its discretions, duties and responsibilities under the Water Affairs Management Law (2020) and abolished the NWARA. Currently, the MEW is the key technical national water governance entity in Afghanistan and the Water Affairs Management Law (2020) is the valid national water law.

Improved Institutions in Afghanistan, Pending Effective Implementation

Afghanistan seems to have improved its institutional capacity for water governance over the last two decades. Successive legislation for water governance has created administrative capacity for water management, but there are several overlaps between the old and new institutions, and severe gaps remain in the implementation of the state's legal and institutional framework for water governance.

Under the Constitution of Afghanistan (2004), natural resources such as water are to be regulated by laws. The Water Act of 2009 and the recent Water Affairs Management Law (2020) have also led to the creation of a range of institutions. The multitude of legal provisions for water in Afghanistan imply that the state is obliged to protect the living environment including rivers and water bodies. In discharging this

Table 8.2. Institutions for water resource governance according to the Water Affairs Management Act (2020). (Adapted from USAID, 2020)[39]

Level	Institution	Role and responsibilities
National	Ministry of Energy and Water (MEW)	Drafting water laws, policies and strategies, analysing the potential of available and accessible water, and watershed management. Preparing plans for water allocation and use. Preparing plans for rehabilitation, development and use of water infrastructure and its execution and supervision. Expansion of hydrological and hydrogeological networks. Creation of a national information bank for water resources for water planning and in due consideration of climate change impacts. Facilitating private investment in the water sector.
	Supreme Council of Water, Land and Environment (SCWLE)	Approving policies and strategies related to the regulation and development of water resources, beneficial use and oversight of its implementation. Approving plans and management of transboundary water. Approving plans of diplomacy for transboundary water negotiations. Approving plans for regulation, allocation, supply, distribution and use of water resources. Deciding on interbasin water transfers. Deciding on transboundary waters.
	National Environmental Protection Agency (NEPA)	Responsible for monitoring surface water quality along with MEW and river basin agencies.
	Ministry of Mines and Petroleum	Responsible for monitoring and managing groundwater quality, groundwater availability and balance studies, and permits for deep wells.
	Ministry of Public Health	Responsible for setting water quality standards based on use, including for drinking, domestic use, irrigation and industrial uses. However, the Ministry of Public Works should do this in co-operation with NEPA, MEW, MAIL and Ministry of Mining and Petroleum.
Sub-national	River Basin Agencies (RBAs), Sub-basin Agencies (SBAs)	Established by NWARA, these institutions are not consistently established in all basins. River basin associations and sub-basin associations develop basin-management and water-use plans in accordance with National Water Policy 2018 and in consultation with river basin councils responsible for surface-water quality monitoring.
	River Basin Councils (RBCs), Sub-basin Councils (SBCs)	MEW is responsible to establish river basin councils and sub-basin councils consisting of water users and government representatives from line ministries. Provide advice on water management strategy and water dispute resolutions and advise on water allocation plans. Monitor water allocations and the protection of water rights.
	Water Users Associations (WUAs), IAs	Water user associations are responsible for managing water use, distribution, operation and maintenance of main canals, and are registered with MEW and NWARA. Irrigation association responsibilities are similar to water user associations but these are registered with the Ministry of Agriculture, Irrigation and Livestock (MAIL). These entities are not consistently established in all river basins.

responsibility, the legal provisions in successive water acts adopt a river basin approach and principles of integrated water resources management to design national water and environmental policies. The uses of water are prioritized, with drinking water and livelihoods the top priority, and with due consideration to informal institutions and local practices. The Water Law (2009) decrees water use to be free but allows water service providers to charge fees for the supply of water and related operating and maintenance costs.

Several institutions were created by the 2009 Water Law (Fig. 8.2). These are the Supreme Council of Water Affairs Management for national co-ordination and then a cascading series of river basin councils, river sub-basin councils and water users associations. While laws confer considerable power to these institutions, river basin councils and river sub-basin councils have been established only for the Kabul and Haridud-Murghab rivers; and whilst water user associations are established in most provinces, they do not appear to be operating in practice (Lackner, 2016).

The Afghan water laws also assign responsibilities related to water to relevant ministries. Importantly, the management and planning for transboundary waters are with the Ministry of Energy and Water with agreements from the Ministry of Foreign Affairs.

The Environment Law (2007) of Afghanistan regulates the use and consumption of water by prohibiting the discharge of effluents into the rivers or waters of Afghanistan without a permit issued by the Environmental Protection Agency. The law requires the agency to establish standards for the discharge of effluents into water. However, these responsibilities tend to overlap with those prescribed by the Water Law (2009) in as much as the latter also allocates the responsibility to establish water quality standards to the Ministry of Agriculture, Irrigation and Livestock, the Ministry of Urban Development and the Ministry of Rural Rehabilitation and Development.

Despite the range of legal provisions for water governance in Afghanistan, water in Afghanistan continues to be governed through a complex interplay of formal and informal rules. While the Water Law (2009) provides an overarching legal and institutional framework for the comprehensive management of water resources, it has not been fully implemented throughout the country, according to a 2015 UN Study (Lackner, 2016). Most farmers still prefer to depend on tribal customary law, which is administered through the *mirabs* or local village masters.

The change in political governance in Afghanistan in August 2021 is likely to halt or even reverse the promises of integrated water resource management and the river basin approach initiated by the Water Law (2009) and reinforced by the Water Affairs Management Act (2020). The implementation of well-meaning water laws is a key challenge on the Afghan side that needs to be addressed, as sole reliance on traditional water governance structures in Afghanistan are unlikely to be able to address problems of water distribution, over-abstraction and pollution in the western transboundary river basins.

The change in political governance in the country has also resulted in a reduction in technical capacity for managing water resources, as experts leave the country or are no longer able to work.[40] Another key concern is the tendency of political regimes to use water and water infrastructure as a weapon such as in the case of Badakhshan district, Kandahar province and Herat. A particularly worrisome issue is Taliban control of the Dahla Dam and their attack on the Afghanistan–India Friendship Dam (Hessami and Ghafoori, 2021). The tendency to target dams as symbolic objects of water infrastructure could threaten prospects of collaboration between Afghanistan and Pakistan on the western transboundary river basins.

Pakistan Water Laws and Institutions

Constitutional provisions

The legal and institutional architecture for water governance on the western side of Pakistan's transboundary river basins rests on a mix of federal laws and provincial legislation of the province of KPK, the only Pakistani province through which the Kabul, Kurram and Gomal rivers flow. An overview of the Constitution of the Islamic Republic of Pakistan provides an insight into the legislative and policy domains of the federation *vis-à-vis* the provinces with respect to water. The Constitution of the Islamic Republic of Pakistan (1973) establishes a three-tier form of government, with a federation consisting of a president, a parliament (comprising a national assembly and senate), a prime minister and the Supreme Court, along with provincial federal units with their own governors, provincial assemblies, chief ministers, provincial high courts and sub-provincial local governments.

Subsequent to the Constitution (Eighteenth Amendment) Act 2010, the parliament is empowered to make laws over subjects enumerated in the Federal Legislative List set out in the Fourth Schedule to the Constitution and provincial assemblies have the power to make laws over subjects not enumerated in the List.[41] Similarly, the Constitution extends the executive authority of the federation to those subjects over which it can make laws[42] and the executive authority of the provinces to those subjects on which it can make laws.[43] Water is not on the Federal Legislative List, making water a subject of provincial legislative and executive competence. Water appears less than ten times in the text of the Constitution of Pakistan, but it is from here we can ascertain the roles of the federation and provinces regarding water.

The Constitution establishes the Council of Common Interests, an apex federal body comprised of the prime minister as chairperson, the provincial chief ministers and three members of the federal government, who have been mandated to formulate and regulate policies in relation to matters enumerated in Part II of the Federal Legislative List. Part II establishes the Water and Power Development Authority under the Pakistan Water and Power Development Authority Act 1958 (the WAPDA Act) to prepare comprehensive plans for the development and use of the water and power resources of Pakistan.[44] In this manner, though water is a provincial subject, the Council of Common Interests may set and regulate policy on the water resources of Pakistan through WAPDA.

In addition, Article 155 of the Constitution gives the Council of Common Interests the power and authority to hear complaints as to interference with water supplies if the interests of the provinces, federally administered tribal areas or capital territory are being affected by (i) any executive act or legislation taken or passed, or proposed to be taken or passed; or (ii) the failure of any authority to exercise any of its powers with respect to the use, distribution or control over water from that source.

In such cases, the federal government or the provincial government concerned may make a complaint in writing to the Council.

Upon receiving a complaint, the Council shall, after having considered the matter, either give its decision or request the president to appoint a commission consisting of such persons having special knowledge and experience in irrigation, engineering, administration, finance or law, as he may think fit, hereinafter referred to as the Commission. After considering the report and supplementary report of the Commission, it shall record its decisions on all matters referred to the Commission.

It can be observed that this constitutional platform is not automatically available. Only the federal or provincial governments may refer a matter to the Council of Common Interests, and if the conditions stipulated in Article 155(1)(a) and (b) are met. However, interprovincial politics, and difficulties in convening or setting the agenda of the Council of Common Interests make it an unreliable platform for adjudication of such complaints. Nevertheless, the Constitution attempts to ensure that, if an adjudication on water supplies does take place, it will be based on experience and not politics.

Federal legislation

Easements Act 1882

An easement is a type of legal right that originated as part of the English Common Law of real property and was introduced to South Asia during the colonial period. An easement is a right in immovable property and is defined as 'a right which the owner or occupier of certain land possesses, as such, for the beneficial enjoyment of that land, to do and continue to do something, or to prevent and continue to prevent something being done, in or upon, or in respect of certain other land not his own'.[45] It is a peculiar right in as much as it vests the owner of the land with rights over another's land in certain circumstances.

In Pakistan, the Easements Act 1882 is particularly relevant for the extraction of groundwater. According to the Act, owners of land had the right to 'collect and dispose within his own limits of all water under the land which does not pass in a defined channel and all water on its surface which does not pass in a defined channel' (Easements Act 1882, Section 7).

Despite the provision in various federal and provincial laws with regard to governance of groundwater, in practice the Easements Act 1882 governs groundwater extraction in the country to date. The Act allows landowners to effectively collect and dispose of underground water as long as it is not a part of a public irrigation network.

However, nothing in the Act derogates from 'Any right of the Government to regulate the collection, retention and distribution of the water of rivers and streams flowing in natural channels, and of natural lakes and ponds, or of the water flowing, collected, contained or distributed in or by any channel or any work constructed at the public expense for irrigation'.[46] Easements, therefore, are property rights that can be exercised between private parties, but not against the government in relation to water that flows naturally, or is found in nature or is part of the public irrigation network.

The Khyber Pakhtunkhwa Water Act 2020 has provisions that give the provincial government control over groundwater extraction and require users to obtain licences for groundwater use. However, these provisions are not currently implemented, hence large-scale groundwater abstraction is taking place unrestrained.

Water and Power Development Authority (WAPDA) Act 1958

The Act provides for 'the unified and coordinated development of the water and power resources of Pakistan'.[47] The Act establishes the WAPDA and gives it the general power and duty to prepare comprehensive plans for the development and utilization of water and power resources including the power to frame schemes providing for irrigation, water supply and drainage and the recreational use of water, flood control and inland navigation.[48] All schemes prepared by WAPDA must be submitted for approval to the Government of Pakistan. The Act also confers onto the Authority the control over underground water resources of any region in a province, subject to other laws.[49] However, before WAPDA exercises any control over underground water resources, the area over which and the extent to which control is intended to be exercised is to be notified by the government.[50] In practice, WAPDA does not take responsibility for groundwater governance.

The jurisdiction of the Council of Common Interests to set policy in relation to the WAPDA and the composition of the Council provide a balance of federal planning, co-ordination and provincial political consent in the framing of any water schemes or exercise of any control by the Authority.

Federal Government Rules of Business 1973

Articles 90 and 99 of the Constitution confer onto the federal government the power to make rules for the allocation and transaction of its business. These rules are in exercise of that power and provide for the constitution and composition of ministries and divisions within the government, the allocation of business therein, and inter-ministerial and cabinet proceedings.[51]

Rule 13 provides that the Foreign Affairs Division be consulted on all matters which affect the foreign policy of Pakistan or the conduct of its foreign relations. Below is a list of ministries and divisions responsible for water issues:

i. Housing and Works Division: physical planning and human settlements including water supply, sewerage and drainage
ii. Water Resources Division: matters relating to the development of water and power resources in the country, Indus Waters Treaty, 1960, Water and Power Development Authority, Indus Rivers System Authority and the Pakistan Transborder Water Organization.

Water Apportionment Accord (1991)

Pakistan's Water Apportionment Accord (1991) is a landmark agreement executed by the chief ministers of the four provinces of Pakistan in March 1991 and relates to the apportionment of the waters of the Indus river. For the implementation of the Accord, the need to establish an Indus River System Authority was accepted.[52] The allocation of each province is set out in paragraph 2 of the Accord (Table 8.3).

The Accord recognizes the need for certain minimum outflow to the sea below the Kotri Barrage (environmental flows necessary to maintain the ecosystem of the Indus Delta) and agreed to conduct studies to establish the optimum level of release. The province of Sindh maintained the optimum level was 12.3 GM.

The apportionment of water in the Accord assumes flows of 144 GM^3/year in the system to be apportioned in the manner provided in paragraph 2. However, in the absence of additional storage capacity, these projected flows have never been reached. As a result, the province of Punjab has argued that water should be apportioned as per the historical uses.[53] The province of Sindh (a lower riparian on the Indus), however, maintains that allocations should be according to the percentages in paragraph 2 of the Accord (Condon *et al.*, 2014). At present, IRSA uses a three-tier method to distribute water. The method consists of three portions of the total water predicted to be available in the system in the upcoming year.[54]

Indus Rivers Systems Authority Act 1992

The Act establishes the Indus Rivers System Authority (IRSA) 'for regulating and monitoring the distribution of water sources of the Indus River in accordance with the Apportionment of the Waters of the Indus Rivers System between the Province signed by the provinces on 16 March 1991 and approved by the Council on 21 March 1991'.

Table 8.3. Allocations by province (billion cubic metres (bcm) per year).

Province	Season: Kharif	Season: Rabi	Total
Punjab	45.72	23.27	69.00
Sindh	41.86	18.28	60.14
NWFP (A)	4.29	2.83	7.13
NWFP (B) civil canals	2.2	1.48	3.7
Balochistan	3.51	1.25	4.77
Total	97.61	47.13	144.74

The IRSA is responsible for establishing the basis for regulation and distribution of surface waters amongst the provinces according to the allocations and polices spelled out in the 1991 Accord.[55] It is also in charge of reviewing and specifying river and reservoir operation patterns and periodically reviews the system of such operations. In addition, IRSA is responsible to co-ordinate and regulate the activities of the Water and Power Development Authority, the exchange of data between the provinces in connection with the gauging and recording of surface flows. IRSA's mandate includes the settlement of any questions that may arise between two or more provinces in respect of distribution of river and reservoir waters.

Section 8(3) of the Act gives provincial governments and WAPDA the opportunity, if aggrieved by any decision of IRSA, to make a reference to the Council of Common Interests. However, it is pointed out that under Article 155(1) of the Constitution, only the federal or provincial governments may make complaints to the Council. It is unclear, therefore, whether the Council of Common Interests would entertain a complaint against IRSA made by WAPDA.

National Water Policy 2018

The National Water Policy 2018 derives from amendments of 2017 to the Rules of Business 1973 of the Executive Authority. Under the Rules of Business 1973, the Ministry of Water Resources has authority over matters related to water resources development in the country. In addition, the Council of Common Interests has a remit over policies related to groundwater, hydropower development and other matters covered under the WAPDA Act of 1958.

The National Water Policy 2018 is a broad policy framework by the federal government 'on the basis of which provincial governments can formulate their respective master plans and projects for water conservation, water development,

Table 8.4. Responsibilities of federal government for water resource governance on Pakistan's side.

Legislation	Institution responsible
Constitution of Pakistan 1973	Council of Common Interests
Easements Act 1882	No formal institution
WAPDA Act 1958	Water and Power Development Authority
Indus Apportionment Accord (1991)	Indus River System Authority Council of Common Interests
National Water Policy 2018	Ministry of Water Resources

and water management'.[56] Being the country's first national water policy, this document is commended as a historic achievement by many. However, in terms of its content, the policy is loosely focused on principles of integrated water resource management for sustainable management of Pakistan's water resources, without an action plan for implementation.

It is pertinent to note that Pakistan's National Water Policy 2018 only acknowledges and refers to the Indus Water Treaty and its transboundary relationship with India in Section 9. Pakistan's reliance on the waters of the western transboundary river basins and the need for an effective transboundary water-sharing agreement with Afghanistan has not been addressed in the policy document.[57] A list of institutions responsible at the federal level for water resources governance in Pakistan is presented in Table 8.4.

Water Laws in Khyber Pakhtunkhwa Province

Khyber Pakhtunkhwa Canal and Drainage Act 1873

The Act provides for the regulation of irrigation, navigation and drainage and states that 'throughout the territories to which this Act extends, the provincial government is entitled to the use and control for public purposes the water of all rivers and streams flowing in natural channels and of all lakes, sub-soil water and other natural collections of still water'.[58]

The Act relates to the application of water for public purposes, the construction and maintenance of works, the supply of water, water rates, canal navigation, drainage, obtaining labour for canal, and drainage works.

Section 5 of the Act gives the provincial government authority to declare the water of any river or stream flowing in a natural channel or of any lake or other natural collection of still water to be applied or used for the 'purpose of any existing or projected canal or drainage work' and sets out the process by which compensation is to be paid to persons deemed affected by such a declaration.

While this Act is still in force, the exercise of its powers has been transferred to the Khyber Pakhtunkhwa Irrigation and Drainage Authority in accordance with the Khyber Pakhtunkhwa Irrigation and Drainage Authority Act 1997 (Section 3.3.3).

Khyber Pakhtunkhwa Rules of Business 1985

Article 139(3) confers on provincial governments the power to make rules for the allocation and transaction of its business. These rules are made in exercise of this power and provide for the constitution and composition of departments within the Government of KPK, the allocation of business therein and interdepartmental and cabinet proceedings.

Rule 11 prohibits departments from authorizing any orders that may affect the finances of the province or which involve relinquishment, remission or assignment of water power rights without previously consulting the Finance Department. According to this Act, the list of departments responsible for water issues are:

i. Environment Department: forest watershed management and management and development of public waters

Fig. 8.3. Discussion on water distribution among water managers.

ii. Irrigation Department: river surveys, construction and maintenance of canals, storage of water and construction of water reservoirs, flood control schemes and administration of the Canal and Drainage Act, 1873
iii. Local Government, Elections and Rural Development Department: water supply and sewerage schemes of local governments
iv. Public Health Engineering Department: drinking-water supply schemes, levy and collection of fees for supply of drinking water and sanitation, and wastewater disposal projects
v. Environment Department: watershed management.

Khyber Pakhtunkhwa Irrigation and Drainage Authority Act 1997

The KPK Irrigation and Drainage Authority Act sought to 'transform the provincial Irrigation Department, which hitherto was responsible for the implementation of the Canal and Drainage Act, 1873 into an autonomous authority and to establish area water boards and farmers' organizations towards development and management of the irrigation, drainage and flood control infrastructures in the province'.[59]

The Act establishes the Khyber Pakhtunkhwa Irrigation and Drainage Authority and aims to transfer the work and responsibilities of the Irrigation Department of the Government of Khyber Pakhtunkhwa, including its employees, to the newly established Authority. This meant that the Authority would exercise all the powers under the Canal and Drainage Act 1873, to fix irrigation rates with area water boards, to operate and maintain irrigation and drainage infrastructure in KPK except private and civil canals, as well as co-ordinate surface and groundwater quantity and quality data from KPK.[60]

Notwithstanding the enthusiasm with which the KPK Irrigation and Drainage

Authority was created, it failed to take on the mandate of the KPK Irrigation Department. Initially, some area water boards and farmer organizations were created under the newly established authority. However, these were later disbanded. While the KPK Irrigation and Drainage Authority Act has not been formally repealed by the KPK government, in practice the Authority is not functional.[61]

Khyber Pakhtunkhwa Forest Ordinance 2002

The Ordinance provides for the consolidation and amendment of laws relating to the protection, conservation and sustainable development of forests and other renewable natural resources. It grants the provincial government the power to declare different types of forests (reserved, protected, *guzara*, village etc.) and sets out the manner in which such forests are to be regulated, protected and conserved.

Section 25 of the Ordinance allows a forest officer, with the previous sanction of the provincial government, to 'stop any public or private way or watercourse in a reserved forest provided that a substitute for the way or watercourse so stopped already exists'. Polluting water in reserved forest is prohibited.[62] Section 36 of the Ordinance allows a forest officer, with the previous sanction of the provincial government, to declare a *guzara* forest and to specify measures for the maintenance of water supply in springs, rivers, tanks and reservoirs.[63] Polluting water in a *guzara* forest is prohibited.[64]

Khyber Pakhtunkhwa Integrated Water Resources Management Board Ordinance 2002

The Ordinance establishes the Khyber Pakhtunkhwa Integrated Water Resources Management Board with the Chief Minister of the Province as Chair and some 18 members. Of these, seven are to be representatives of the Kabul and Indus rivers, and the Kohat, Kurram, Gomal, Chitral and Siran river valleys. The remainder are *ex officio* officials of the provincial government. The functions of the board include conducting research and allocating water resources for consumptive and non-consumptive use, to develop policies to protect the contamination and water resources and specify wastewater quality in public drains.

In practice, the board is not functional and its functions are not being subsumed within the KPK Water Resources Commission or KPK Water Resources Regulatory Authority.

Khyber Pakhtunkhwa River Protection Ordinance 2002

The Ordinance provides for the protection of the aquatic ecology, water quality, and economic and environmental value of the rivers and their tributaries in Khyber Pakhtunkhwa Province.

Section 3 prohibits construction or 'any other developmental work' within 200 feet beyond the high-water limit on either side of rivers or their tributaries; the deposit or release or any substance into rivers or tributaries in excess of the National Environmental Quality Standards; and the disposal of any hazardous waste or substances notified by the provincial government into the rivers or their tributaries. Section 4 allows the provincial government to prepare land-use and zoning plans and make building control regulations to control developmental activities and construction for a specific river or all rivers and their tributaries, and construction of buildings in the catchment area of any river.

The Ordinance provides for an authorized officer to enter buildings and property for the purposes of inspection and compliance with the provisions of the Ordinance.[65] Such authorized officer has the power to stop any activity with respect to which an office appears to have been committed under the Ordinance, and those found contravening the Ordinance shall face punishment of imprisonment of up to six months or with a fine of up to PKR 500,000 or both.[66]

Khyber Pakhtunkhwa Local Government Act 2013

The Act provides for the regulation of local government institutions in the province of Khyber Pakhtunkhwa and establishes local governments at the district, tehsil (township) and village levels.

The Act defines 'municipal services' as including intra-city networks of water supply, sanitation and removal and disposal of sullage, sewers and storm water.[67]

The district councils have the function to review the development of integrated water reservoirs, water sources, treatment plants, drainage, liquid and solid waste disposal, sanitation and other municipal services. At the tehsil level, tehsil municipal administrations in urban tehsils have the power to impose and collect taxes on, and levy charges for, the development of water supply and drainage it carries out.[68] At the village level, it is the function of the village council to improve water supply sources, maintain water distribution systems and take measures to prevent the contamination of water.[69] The *nazim* (chief) of a village has the responsibility to report any breach of public water courses.

Khyber Pakhtunkhwa Environmental Protection Act 2014

The Act provides for the protection, conservation, rehabilitation and improvement of the environment, monitoring, prevention and control of pollution and the promotion of sustainable development.[70] The Act defines the environment to include 'air, water and land' and its interrelationship with several other factors, including the ecosystem and ecological relationships and all social and economic conditions affecting community life.[71]

The Act establishes the Khyber Pakhtunkhwa Environmental Protection Council with functions and powers to approve the Khyber Pakhtunkhwa Environmental Quality Standards and provide guidelines for the generation of renewable and non-renewable resources, solid waste and water and sanitation. In this manner, the Council may set standards and guidelines with respect to water.

The Act also establishes the Khyber Pakhtunkhwa Environmental Protection Agency, the powers and functions of which are to be exercised by a director-general or by staff specially delegated by the director-general. The functions of the agency are to prepare, revise, establish and ensure enforcement of the Khyber Pakhtunkhwa Environmental Quality Standards approved by the council and to establish the standards for the quality of ambient air, water and land by notification.[72]

Section 12 of the Act stipulates that the provincial government may be asked to carry out strategic environmental assessments of plans and policies, including water use management and prevention of water pollution into rivers. The Act defines a strategic environmental assessment as a systematic analysis that ensures the principles of sustainable development are integrated into the development of provincial government policies, plans or programmes.[73]

The Act prohibits the discharge of emissions in excess of the Environmental Quality Standards set by the Council. Projects are subject to an initial environmental examination and environment impact assessment process. The agency may pass an Environment Protection Order (Section 17) where satisfied that a discharge or emission has or is about to take place or a violation of the Act has or is about to take place.

Khyber Pakhtunkhwa Water Act 2020

The Government of KPK recently approved the Khyber Pakhtunkhwa Water Act 2020 (KPWA, 2020) through the provincial assembly on 24 July 2020. The Act commits the GoKP to establish the KP Water Resources Commission (KP-WRC) and KP Water Resources Regulatory Authority (KP-WRRA) within six months after the passage of this Act. This is a commendable shift as the preamble states, 'comprehensively manage and regulate' the province's water resources 'in the interest of conservation and sustainability' (KP Water Act, 2020). The first meeting of the KP-WRC was held in 2021. It is a 22-member body headed by the chief minister of the province including public and private representation. KP-WRC can further include co-opted members as it may deem appropriate. KP-WRRA is a 12-member body headed by the additional Chief Secretary of KPK. The Acting Director-General for the KP-WRC is the Secretary of Irrigation and the Acting Director-General for the KP-WRRA is Chief Engineer North. Section (4) of the Act clearly mentions that 'Nothing shall be construed as relieving any water service provider of the obligations to develop water resources for the purpose of any duty imposed on it by virtue of the Act'. This means the existing laws and regulations will hold and the obligations of the departments under those laws will stand even after the passage of this Act. The Rules of Business for both KP-WRC and KP-WRRA are currently being drafted and a plan is being put in place to fund its operations and staffing.

For the most part, this legislation does not address some of the key deficits in institutions and processes adopted to date, particularly with regard to farmer institutions. The Act is not limited to the provision and management of irrigation water but to the management of the province's overall water supplies, drainage, sewerage and other patterns of use. Under the Act, the Commission will have the power to allocate water across seven sectors and purposes (Sec. 43(2)), including for ecological purposes, and will be free to work out the ways and methods by which it undertakes reallocations (Sec. 4 (a)(b)). Overall, under Sec. 4 (c) it will be empowered to work out what it deems the 'proper' use of the province's water resources. The concept of 'proper' will shift depending on the perspective of the party advocating a particular allocation and so it will be crucial for GoKP to lay out at the earliest its concept of 'proper' as well as the mechanism the Commission will use in developing its view of the matter. This is especially important as the Commission will determine the allocations between sectors annually, and given the criteria it will use, the allocations between existing uses may shift. Given the stakes of changing allocations between sectors, it is necessary that the criteria for shifts are transparent and available at the earliest time such that parties and sectors can make informed decisions. Given that the Act's proposed model is one in which the Commission or the Authority (see First Schedule for Sec. 43) will authorize service providers for various sectors and purposes, perhaps the process of taking account of competing purposes can be evaluated from the start at the point of authorization of the service-providing entity.

The KPK Irrigation and Drainage Authority Act of 1997 also envisaged farmers organized in area water boards, farmer organizations, and water user associations as service providers to their community. It is pertinent for the KP-WRRA to re-examine the implementation of the PIDA reforms in KP province and devise a mechanism whereby farmers' participation may be ensured. It is in the mandate of KP-WRRA to approve, determine or revise tariffs for water as well as for service providers. This has been highlighted as an issue in the implementation of the PIDA reforms. Perhaps KP-WRRA can play an instrumental role in bringing clarity to the roles and responsibilities of the existing PIDA and inclusion of stakeholder farmers in the decision-making process.

In Pakistan, the irrigation management transfer and decentralization of the irrigation reform process over the last four decades has been built on the recognition that enabling and preserving farmer participation and involvement in self-governance, particularly in the operation and maintenance of the irrigation network, is a valuable public good. However, any reform effort must contend with existing institutional structures and political economy influences as they operate in practice versus how an idealized design may operate in theory.

The KPK Water Act of 2020 does not address or even mention the issue of shared water resources between Afghanistan and Pakistan. Nevertheless, it is hoped that KPK-WRC will be able to advise the federal government on the allocations of water coming from the western transboundary rivers shared between Afghanistan and Pakistan. This will provide the KPK government with an avenue to work with the federal government and create a co-operation mechanism on the Indus waters coming from the three rivers shared between Afghanistan and Pakistan that fall within the jurisdiction of the Government of KPK. Table 8.5 enlists the responsibilities of the KPK government for water resource governance.

The KPK Water Act (2020) has placed matters related to groundwater extraction and regulation under the KPK Water Resources Commission and the KPK Water Resources Authority. According to the new Act, all users are required to obtain a

Table 8.5. Responsibilities of KPK government for water resource governance.

Legislation	Institution responsible
KPK Canal and Drainage Act 1873	KPK Irrigation Department
KPK Rules of Business 1985	Environment Department KPK Irrigation Department Local government Public Health and Engineering Department
KPK Irrigation and Drainage Authority Act 1997	KPK Irrigation and Drainage Authority
KPK Forest Ordinance 2002	Provincial government
KPK Integrated Water Resource Management Board Ordinance 2002	KPK Integrated Water Resource Management Board
KPK River Protection Ordinance 2002	Provincial government
KPK Local Government Act 2013	Local governments at the district, tehsil and village levels
KPK Environmental Protection Act 2014	KPK Environmental Protection Council KPK Environmental Protection Agency
KPK Water Act 2020	KPK Water Resources Commission KPK Water Resources Authority

licence for groundwater extraction (KPK Water Act, Ch. 8). The charges for the water used are to be regulated by the Water Resources Authority. The creation of the KPK Water Resources Commission is promising more integrated management of groundwater and better control over abstraction. However, almost two years after the approval of the KPK Water Act of 2020, its implementation is still being planned.

The Challenges of Federalism in Water Governance in Pakistan

The above analysis of Pakistan's legal provisions for water resource governance at the federal and provincial levels reveals the complexity of water management posed by federalism. The range of federal and provincial institutions for water governance often results in overlaps in institutional functions or gaps in the governance of water.

As discussed earlier, the control and governance of water, as per the Constitution, lies with the provinces. In contrast to this general rule is the exception of the Council of Common Interests, the Water and Power Development Authority and the Indus River System Authority. The Council of Common Interests can set the policy of the Water and Power Development Authority, which, in turn, by law, has the power to develop hydropower resources and underground water resources in any region of the country. To this extent, therefore, the federal government can still set policy on the subject of water. The federal Rules of Business support this proposition, with the Ministry of Water and Power responsible for the development of water and power resources, the Indus Waters Treaty 1960, the Water and Power Development Authority, the Indus River System Authority, as well as the Pakistan Transborder Water Organization.

The Indus River System Authority regulates the supply of surface water in accordance with the Water Accord. However, it is not concerned with groundwater. Abstractions from the Kabul river in Afghanistan will result in reduced flows in the Kabul river in Pakistan and, in turn, in the manner of allocating the shares of the provinces to the limited supply of water in the Indus river. IRSA does have the statutory authority to direct the Water and Power Development Authority, but it is doubtful such authority would be recognized if it is in conflict with a policy decision of the Council of Common Interests on the same subject.

Federal law in Pakistan also recognizes a number of personal rights in water arising from the ownership and use of property. These individual water rights are effective against private parties but cannot be enforced against water being diverted or used by the government.

At the provincial level, legal and institutional provisions for water governance are well developed but face issues of overlapping jurisdictions and gaps in implementation. Often, new laws have superseded old laws, but it has not been stated that the old laws have been repealed. The control of water resources in the province of KPK is the responsibility of several departments created by old and new laws. While the Irrigation and Drainage Authority was created for the equitable and efficient use of water resources through the area water boards and farmers' organizations, in practice the Authority is not functional. There is a Khyber Pakhtunkhwa Integrated Water Resources Management Board with considerable powers that, if collectively exercised, could help with integrated water resource management in the province. However, the Board has largely been defunct in recent times. The KPK Water Resources Commission has been mandated under the KPK Water Act of 2020 to look after many of the functions that were originally specified under the KPK Integrated Water Resources Management Board.

The recently formed KPK Water Resources Commission and the KPK Water Resources Regulatory Authority try to embody a holistic approach towards the management of water resources in the province. However, since these institutions have been constituted recently, their efficacy with regard to water governance remains to be seen. The relationship between the recent KPK Water Act 2020 and the newly constituted KPK Water Resources Commission with existing institutions and laws related to water governance in KPK is also unclear.

In line with the challenges posed by federalism for water governance in Pakistan, groundwater governance is a key challenge that needs to be addressed. The National Water Policy 2018 called for the creation of a groundwater authority in Islamabad as well as water authorities within the provinces. However, more than three years into the policy's enactment, there has been little progress on its promise of sustainable groundwater governance. While IRSA and the Council of Common Interests have a mandate over interprovincial water distribution, groundwater governance is not completely or clearly allocated to any institution. The KPK Water Act (2020) places all provincial water resources, including groundwater, under the control of the KPK Water Commission and the KPK Water Authority. This provides an opportunity for integrated management of groundwater resources, yet the challenge lies in the implementation of the Act.

Strengthening, Co-ordinating and Engaging Internal Governance for Improved Transboundary Water Co-operation

In recent years, there has been an increased effort within both Afghanistan and Pakistan to overhaul water governance. On the Pakistan side, there has been a progressive move towards better water governance at the federal and provincial levels exemplified by the country's efforts in enacting its first ever National Water Policy, as well as in the recent KPK Water Act of 2020. On the Afghanistan side, the Afghanistan Water Law (2009) and the Water Affairs Management Law (2020) signal the country's commitment to integrated water resource management and a river basin approach towards the governance of its water resources.

There are considerable similarities in the composition of internal water governance mechanisms within Afghanistan and Pakistan. While both countries have a range of formal national institutions and legal mechanisms for water governance, there is considerable reliance on local informal institutions for managing water. While the reliance on informal institutions is more pronounced in Afghanistan, customary water management is also prevalent in the tribal belt in the province of KPK in Pakistan.

Both Afghan and Pakistani water laws are primarily focused on surface water. While there are provisions in the laws of both countries that regulate the use and control of groundwater, neither country has any reliable data on the share of aquifers fed by the Kabul, Kurram and Gomal rivers. International transboundary water law emphasizes the need for co-basin states to treat ground and surface water conjunctively while negotiating any transboundary water sharing arrangement.

There is a need for improved co-ordination of national water governance policies in Afghanistan and Pakistan as stepping-stones to effective transboundary co-operation in the shared basins. As it stands, Afghan water governance is centralized with a river basin approach, deployment of river basin councils and sub-basin councils in line with the principles of integrated water resource governance (Shams and Muhammad, 2023). Water is a provincial subject in Pakistan, with the exception of the federal authority for the Council of Common Interests, WAPDA and IRSA to intervene in relevant matters. Federalism in Pakistan's water governance poses critical challenges to effective water governance both within the country and in its transboundary relationship with Afghanistan.

Afghan water law differs from Pakistani water law in that it would explicitly approach any arrangement on shared rivers using the river basin approach. This is distinct from the transboundary arrangements reached earlier in the Helmand River Treaty and the Indus Water Treaty, neither of which can be said to have employed a river basin approach. The former guarantees Iran a share of water from the Helmand and the latter divides the waters of the major tributaries of the river Indus between India and Pakistan but not between Afghanistan and Pakistan. Pakistani law is silent on how it would approach sharing water and has so far been accustomed to a division of water resources approach as evidenced in its relationship with India over the Indus Water Treaty since 1960. The Irrigation and Drainage Authority in KPK is required to ensure 'optimal utilization' of water on an 'equitable and efficient basis' but is not functional in practice. The recently established KPK Water Commission is silent on the issue of transboundary waters of the western transboundary river basins.

While efforts have been made to overhaul formal legal provisions and internal water governance in both countries, these have been stymied by implementation gaps and political economy issues. On Afghanistan's side, the recent implementation of well-meaning reforms has been stymied by a change in the political regime. On the Pakistan side, the National Water Policy 2018 still awaits implementation almost three years after its enactment. Similarly, the relationship of the KPK Water Commission with existing institutions for water governance in the province is unclear, as well as the Commission's mandate to contribute to any transboundary water sharing agreements on the Kabul, Kurram and Gomal river basins.

Engaging multilevel stakeholders for improved co-ordination of water policies between Afghanistan and Pakistan can pave the way towards successful transboundary basin management in the Kabul, Kurram and Gomal river basins. There is an active debate in Pakistan to establish a mechanism of water sharing between Afghanistan and Pakistan on the western transboundary river basins, particularly the Kabul river basin (e.g. Basin, 2010; Shah and Nafees, 2020). In this regard, the role of the Pakistan Commission for Indus Waters (PCIW) becomes extremely important. There are some major changes in the PCIW which are currently being considered to address the issue of water sharing between Afghanistan and Pakistan. The Pakistan Commission for Indus Waters (PCIW) is part of the Permanent Indus Commission, which is a bilateral commission consisting of officials from India and Pakistan and was created to implement and manage the goals and objectives of the Indus Waters Treaty. The Commission maintains and exchanges data and facilitates co-operation between the two countries. Recently, its mandate has been proposed to be expanded to cover the other transboundary tributaries of the Indus with a change in name to the Pakistan Commission for Indus and Transboundary Waters. Under the newly proposed organization, three cells operate under the Pakistan Commission for Indus and Transboundary Waters. The Indus Water Cell, Research Cell and the Kabul River (Afghan) Cell. All matters pertaining to the western transboundary river basins, particularly the Kabul river, will be dealt with by the Afghan Cell. The cell will be supported by a joint commissioner, a deputy commissioner and an assistant commissioner dedicated to this cell. The summary for the new organization has been approved and the human resource required to take on these roles is being recruited. It is expected that once these cells are fully resourced, it will allow the Commission to develop a more comprehensive strategy to deal with the affairs of the western transboundary river basins shared between Afghanistan and Pakistan.

Any efforts at transboundary co-operation will need an appreciation of the multi-scalar nature of transboundary basin management and the need to engage national, subnational and local institutions and stakeholders. The authority to engage in bilateral negotiations on the Kabul, Kurram or Gomal rivers would have to originate from the governments of both countries through their respective ministries of foreign affairs, before any agreement could be reached on water sharing or allocation in any transboundary river. This provides an opportunity for reliable data about water flow in the rivers and their tributaries to be made available to both sides. The collection of such data on the Afghanistan side would primarily be the responsibility of the Ministry of Energy and Water and the Ministry of Agriculture, Irrigation and Livestock. The collection of such data in Pakistan would be the responsibility of the Khyber Pakhtunkhwa Irrigation and Drainage Authority, the Water and Power Development Authority and the Indus River System Authority.

How water is managed at the local, provincial and national levels has an important bearing on the water governance at the shared-basin level. Ultimately, meaningful transboundary co-operation cannot be achieved – even if Afghanistan and Pakistan enter into an agreement on the shared river basins – if internal water governance is not aligned towards achieving the benefits of co-operation (Sadoff and Grey, 2002).

Notes

[1] There were no provincial laws on water that were identified during research for this paper.
[2] Article 2347 of the Afghanistan Civil Code (1977).
[3] Articles 2346–2347, Ibid.
[4] Article 2349, Ibid.
[5] Article 2353, Ibid.
[6] Article 2348, Ibid.
[7] Article 2352, Ibid.
[8] Article 2350, Ibid.
[9] Article 2351, Ibid.
[10] Article 1, Environment Act (2007).
[11] Article 4(10), Ibid.
[12] Article 34(1), Ibid.
[13] Article 34(2), Ibid.
[14] Article 35, Ibid.
[15] Preamble to Afghanistan's Water Law (2009).
[16] Article 2, Ibid.
[17] Article 40, Ibid.
[18] Article 4, Ibid.
[19] Article 5, Ibid.
[20] Article 6, Ibid.
[21] Article 7, Ibid.
[22] Article 8, Ibid
[23] Articles 10 and 11, Ibid.
[24] Article 12(1), Ibid.
[25] Article 12(2), Ibid.
[26] Article 14, Ibid.
[27] Article 18, Ibid.
[28] Article 16, Ibid.
[29] Article 18, Ibid.
[30] Article 23, Ibid.
[31] Article 24, Ibid.

[32] Article 25, Ibid.
[33] Article 27, Ibid.
[34] Article 34, Ibid.
[35] Article 35, Ibid.
[36] Rasooly, 2019
[37] Rasooly, 2019
[38] https://winrock.org/wp-content/uploads/2021/08/Afghanistan_Country_Profile-Final.pdf
[39] https://winrock.org/wp-content/uploads/2021/08/Afghanistan_Country_Profile-Final.pdf
[40] Personal communication (Afghanistan water resource governance specialist)
[41] Article 142 of the Constitution of Pakistan. Note that Article 142(b) allows both parliament and provincial assemblies to make laws with respect to criminal law, criminal procedure and evidence.
[42] Article 97 of the Constitution of Pakistan.
[43] Article 137 of the Constitution of Pakistan.
[44] Section 8(1) of the Water and Power Development Authority Act 1958.
[45] Section 4 of the Easements Act 1882
[46] Section 2(a) of the Easements Act 1882.
[47] Preamble of the Water and Power Development Authority Act, 1958.
[48] Section 8(1) and 8(2) Ibid.
[49] Section 11(1)(i)(a) Ibid.
[50] Section 11(2) Ibid.
[51] The Rules were made in 1973 but most recently updated in 2015 to reflect change post-18th Amendment.
[52] Paragraph 13 of the Water Apportionment Accord (1991). The Indus Rivers System Authority was established under the Indus Rivers Systems Authority Act 1992 (discussed in section on Water Apportionment Accord (1991) above).
[53] See paragraph 14(b), Ibid.
[54] Ibid., p. 23.
[55] Section 8 of the Indus Rivers Systems Authority Act 1992.
[56] National Water Policy 2018, p. 5.
[57] National Water Policy 2018, p. 15.
[58] Preamble to the Khyber Pakhtunkhwa Canal and Drainage Act 1873.
[59] Preamble to the Khyber Pakhtunkhwa Irrigation and Drainage Act 1997.
[60] Section 8, Ibid.
[61] Personal communication, KPK Irrigation Department
[62] Section 33(2)(g) Khyber Pakhtunkhwa Forest Ordinance 2002
[63] Section 36(1)(v) Ibid.
[64] Section 44(g) Ibid.
[65] Section 8 of the Khyber Pakhtunkhwa River Protection Ordinance 2002
[66] Sections 10 and 11, Ibid.
[67] Section 2(r) of the Khyber Pakhtunkhwa Local Government Act 2013
[68] No. 4, Part II, Third Schedule, Ibid.
[69] Section 29 Ibid.
[70] Preamble to the Khyber Pakhtunkhwa Environmental Protection Act 2014
[71] Section 2(r), Ibid.
[72] Sections 6(1)(v), (vi) and (vii) Ibid.
[73] Section 2(eee) Ibid.

References

Basin, K.R. (2010) *Towards Kabul Water Treaty: Managing Shared Water Resources: Policy Issues and Options*. IUCN, Pakistan.

Condon, M., Kriens, D., Lohani, A. and Sattar, E. (2014) Challenge and response in the Indus Basin. *Water Policy* 16(S1), 58–86. DOI: 10.2166/wp.2014.004.

Earle, A. and Neal, M.J. (2017) Inclusive transboundary water governance. In: Karar, E. (ed.) *Freshwater Governance for the 21st Century*. SpringerLink, pp. 145–158.

Hessami, E.B. and Ghafoori, A. (2021) Taliban seize Herat and assault nearby dam that provides water and power to hundreds of thousands of Afghans. Available at: https://www.downtoearth.org.in/blog/water/taliban-seize-herat-and-assault-nearby-dam-that-provides-water-and-power-to-hundreds-of-thousands-of-afghans-78447 (accessed 21 July 2022).

KPWA (2020). Available at: https://kpcode.kp.gov.pk/uploads/The_Khyber_Pakhtunkhwa_Water_Act_2020_Act_No_XXV_of_2020.pdf

Lackner, M. (2016) Water rights: an assessment of Afghanistan's legal framework governing water for agriculture. Available at: file:///C:/Users/Terry%20Erle/Downloads/2016_19_10_water_rights_final_v2%20(1).pdf (accessed 21 July 2022).

Moss, T. and Newig, J. (2010) Multilevel water governance and problems of scale: setting the stage for a broader debate. *Environmental Management* 46(1), 1–6. DOI: 10.1007/s00267-010-9531-1.

Newig, J. and Fritsch, O. (2009) Environmental governance: participatory, multi-level - and effective? *Environmental Policy and Governance* 19(3), 197–214. DOI: 10.1002/eet.509.

Rasooly, N. (2019) Water governance in Afghanistan. MSc thesis, Oregon State University. Available at: file:///C:/Users/Terry%20Erle/Downloads/RasoolyNawid2019%20(3).pdf (accessed 21 July 2022).

Rawlins, J. (2019) Harmonisation of transboundary water governance: advance or align? Available at: https://www.africaportal.org/features/harmonisation-transboundarywater-governance-advance-or-align/ (accessed 21 July 2022).

Sadeqinazhad, F., Atef, S.S. and Amatya, D.M. (2018) Benefit-sharing framework in transboundary river basins: the case of the Eastern Kabul River Basin-Afghanistan. *Central Asian Journal of Water Research* 4(1), 1–18. DOI: 10.29258/CAJWR/2018-RI.v4-1/1-18.eng.

Sadoff, C. and Grey, D. (2002) Beyond the river: the benefits of cooperation on international rivers. *Water Policy* 4(5), 389–403. DOI: 10.1016/S1366-7017(02)00035-1.

Schulz, C., Martin-Ortega, J., Ioris, A.A.R. and Glenk, K. (2017) Applying a 'Value landscapes approach' to conflicts in water governance: the case of the Paraguay-Paraná Waterway. *Ecological Economics* 138, 47–55. DOI: 10.1016/j.ecolecon.2017.03.033.

Shah, W. and Nafees, Md. (2020) Trans-boundary water conflict between Pakistan and Afghanistan and need of a water treaty on the Kabul River. *International Review of Basic and Applied Sciences* 8(4).

Shams, A.K. and Muhammad, N.S. (2023) Towards sustainable transboundary water cooperation between Afghanistan and Pakistan: a case study of Kabul River. *Ain Shams Engineering Journal* 14(2), 101842. DOI: 10.1016/j.asej.2022.101842.

USAID (2020). Available at: https://winrock.org/wp-content/uploads/2021/08/Afghanistan_Country_Profile-Final.pdf

World Bank (2017) *Political Economy Analysis for Transboundary Water Resources Management in Africa: Practical Guidance*. World Bank, Washington, DC. DOI: 10.1596/25824.

Yasin, H.Q., Breadsell, J. and Tahir, M.N. (2020) Climate-water governance: a systematic analysis of the water sector resilience and adaptation to combat climate change in Pakistan. *Water Policy* 23(1), 1–35. DOI: 10.2166/wp.2020.113.

Young, W.J., Anwar, A., Bhatti, T., Borgomeo, E., Davies, S. *et al.* (2019) *Pakistan: Getting More from Water*. World Bank, Pakistan. DOI: 10.1596/31160.

Zeitoun, M., Cascão, A.E., Warner, J., Mirumachi, N., Matthews, N. *et al.* (2016) Transboundary water interaction III: contest and compliance. *International Environmental Agreements* 17(2), 271–294. DOI: 10.1007/s10784-016-9325-x.

9 Co-operation in the Afghanistan–Pakistan River Basins

Jonathan Lautze, Asadullah Meelad and Shakeel Hayat

Chapter Overview

In transboundary waters, co-operative basin management is widely believed to offer benefits over and above those of separate and unilateral management by riparian countries. Nonetheless, co-operation on shared waters does not always exist. The Kabul, Kurram and Gomal basins, shared by Afghanistan and Pakistan, present examples of river basins that lack co-operation despite the benefits that could be accrued. This chapter takes stock of existing co-operation in the Kabul, Kurram and Gomal basins to suggest catalysts for co-operation. The chapter reviews the current status of water co-operation between the two countries and the adverse consequences due to the current status quo. Factors constraining co-operation are then presented, followed by the benefits that can be derived from co-operation and frameworks that can guide its development. Finally, practical incentives for scaling up co-operation, and specific small steps that can be taken, are presented. A particular value-addition of this chapter is a granular distillation of the key points on which co-operation should focus and the mechanisms and modalities that can enable that focus.

Fig. 9.1. Band-e-Amir, water reservoir, Afghanistan. (From: Alex Treadway, ICIMOD)

DOI: 10.1079/9781800622371.0009

Introduction

Why co-operation on shared waters? The rationale for co-operation on shared waters is multipronged. At a practical level, water is widely viewed to be optimally managed according to basin boundaries. When basin boundaries cross international borders, co-operation is thus necessary to realize optimal management that maximizes economic benefits (e.g. hydropower, irrigation) and minimizes risks (e.g. floods, droughts). As such, engagement on transboundary water issues constitutes one avenue through which states advance their national interests. Equally, international waters have been identified as a potential source of conflict, and efforts toward co-operation have been shown to reduce this potential (Yoffe *et al.*, 2003). Reflecting the importance of optimizing basin benefits and avoiding conflict, international conventions such as the UNECE Water Convention (2016) and the UN Watercourse Convention (1997, Article 8) call for co-operation on international watercourses. Furthermore, realizing a transboundary operational arrangement is currently an indicator of progress towards optimal management of a basin and the Sustainable Development Goals Fig. 9.2.

- Application of international water conventions and customary international water law call for co-operation to ensure equitable and sustainable sharing.
- Consistent with best practices in basin management, jointly managing shared basins as units would enable greater realization of benefits and minimization of risks.
- Water co-operation reduces the potential for conflict and promotes peace and security at various levels (e.g. interstate, national and local).

Practical drivers for co-operation are less clear-cut. While basins are indeed best managed collectively, effective basin management often requires water storage infrastructure that may not sufficiently exist in shared Afghan–Pakistani waters.

Fig. 9.2. Group photo of Afghan and Pakistani nationals at a workshop in Tashkent, April 2019 (IWMI).

What level of co-operation is desirable? Existing literature (e.g. Yousaf, 2017; Malik, 2019; Khan *et al.*, 2020; Tariq *et al.*, 2020) and previous chapters of this book have been clear on the need for water co-operation between the two countries. Chapter 4 outlines how improved responses to climate change can benefit from basin management, for example, and Chapter 5 points to opportunities for improved flood control. Water co-operation can, nonetheless, take many forms. In this context, the central issue may be determining the form of co-operation best suited to the geographical and development context, cognizant of other agreements in the basin such as the Indus Water Treaty (1960), and aligned with existing international legal principles and norms. The topic has received some attention. Pervaz and Khan (2014) looked at options for legal frameworks. Vick (2014) highlighted how co-operation can include data exchange and communication and donor engagement. Atef *et al.* (2019) outlined a step-by-step process through which countries could arrive at a framework for benefit-sharing. Panikkar *et al.* (2019) drew attention to, at times, overlooked aspects of co-operation such as community engagement and social learning that may be most suited to this context. However, the existing literature stops short of any clear-cut identification of elements that could form key tenets of Afghan–Pakistani water co-operation, both under the present conditions and in the future.

This chapter. This chapter is focused on water co-operation in three shared basins of Afghanistan and Pakistan: the Kabul, Kurram and Gomal basins. The chapter is oriented towards taking stock of existing context to suggest suitable steps toward and catalysts for co-operation, both now and in the future. Structurally, the chapter reviews the status of water co-operation between the two countries and the adverse consequences due to the current status quo. Factors constraining co-operation are then presented, followed by the benefits that can be derived from co-operation and frameworks that can guide its development. Finally, practical incentives to scaling up co-operation, and specific small steps that can be taken, are presented. A particular value-add of this chapter is the granular distillation of the key points on which co-operation should focus, as well as the mechanisms and modalities that can enable that focus.

Status of Water (Non)Co-operation

Co-operation on Indus tributaries was unnecessary before the Common Era and through the Islamic era (n.d.–1857). Until the mid-19th century, relatively low levels of water use and the absence of international boundaries did not require transboundary basin management. In Afghanistan, water governance was traditionally undertaken by community-based management structures with elected or, more often, selected water masters (*mirabs*) who supervised the construction and maintenance of water infrastructure, the implementation of rules and the peaceful resolution of conflicts (McCarthy and Mustafa, 2014). Water allocation or rights to access water in the region are historically linked to land ownership, the size of that land and the level of contribution to the construction and maintenance of water infrastructures (Qureshi, 2002; ALEP, 2011). Although customary practices provide a solid and important foundation for shaping indigenous water governance, they are often subject to elite capture as customary or informal structures are not always inclusive (Weinthal *et al.*, 2014). Water law in Afghanistan today is a complex mixture of customary practices, Islamic rules and modern principles (Hessami, 2018). In the past, low levels of water use did not require basin-level co-ordination, and the lack of international borders meant that Indus basins were not transboundary.

Water governance in the colonial era (1857–1947). Freshwater governance in Afghanistan and Pakistan gradually emerged from traditional irrigation practices as well as principles of Islam, which were merged with modern conceptions of water management during the British colonialization of India (Abderrahman, 2000). Indus basin tributaries became transboundary in the colonial era, ultimately formalized by the creation of the 2430 km Durand Line in 1893 to clarify the spheres of influence between Afghanistan and British

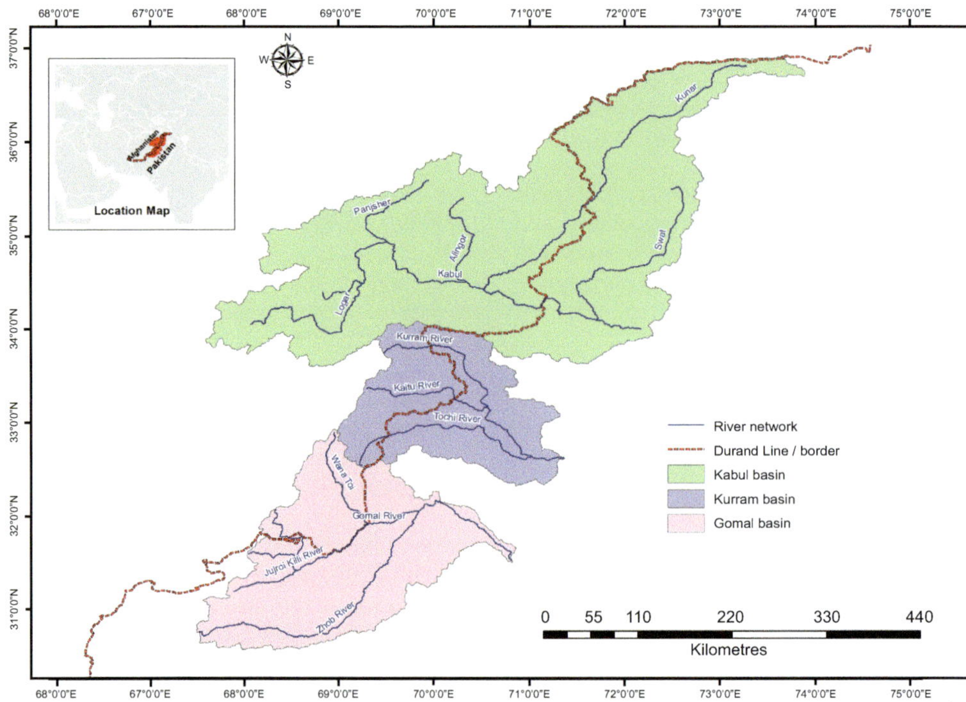

Fig. 9.3. Durand Line. (Adapted from International Water Management Institute)

India and improve bilateral relations, including trade in the region (Omrani, 2009; Kaura, 2017). The Durand Line cuts through three Indus tributaries, politically dividing ethnic groups and internationalizing the Indus basin (Fig. 9.3). Before the Durand Line Agreement, the Kabul, Kurram and Gomal rivers were an integral part of the kingdom of Afghanistan. Upon independence, Pakistan inherited the Durand Line Agreement. However, Kabul neither signed nor ratified a formal agreement on border issues (Janjua, 2009; Biswas, 2013; Yousafzai and Yaqubi, 2017) or shared water resources directly with Islamabad (Ahmad, 2010; Azam, 2015; Kakakhel, 2017).

One colonial era agreement. One transboundary water governance framework was created in the colonial period that applied to the Kabul river, a tributary to the Indus; the treaty between British India and the Afghan government in 1921. Although its reference to water was limited, there was a clear water use mechanism (SIWI, 2015; Saeed *et al.*, 2016). This treaty comprised 14 articles and two schedules and was valid for three years from the date of signing. According to Article 2, the British colonialists agreed to permit the residents of Torkham in Afghanistan (a border village and crossing-point between Afghanistan and Pakistan) to draw water through a pipeline (Article 2). In return, Afghanistan would allow the British officers and tribesman to use the water resources of only the Kabul river for navigation and to sustain their current irrigation rights (Favre and Kamal, 2004).

Water governance practices in the modern era (1947 to the present). There is no co-operative framework between Afghanistan and Pakistan on Indus basin tributaries. Nonetheless, Afghanistan, Pakistan and the international community have made efforts to formalize water sharing between the two neighbours. At least six efforts have been made since 2003.

- *Kabul River Treaty.* In 2003, the federal minister of Pakistan for water and power proposed to Afghanistan's Ministry of Foreign Affairs a possible Kabul river treaty to efficiently and peacefully use shared water resources. However, Afghanistan declined because it lacked the technical and financial capacities to sign a bilateral treaty with Pakistan (Vick, 2014; Kakakhel, 2017). In 2004, Pakistan reiterated an offer for a transboundary agreement over the Kabul river basin that would enable making long-term water availability projections (Kakakhel, 2017).

- *USAID–World Bank Dialogue.* In 2006, USAID actively supported the idea of formally sharing water resources in the Kabul river basin and initiated dialogues between the two neighbours. In 2006, the World Bank also offered to arbitrate a dialogue over water issues between Afghanistan and Pakistan but was not successful as Afghan officials refused to collaborate or share water-related data.
- *Kabul River Basin Management Commission.* In 2011, the World Bank offered to help establish a Kabul River Basin Management Commission (KRBMC) for conflict resolution and joint management. The core purpose of establishing the KRBMC was to enhance collaboration, build institutional capacity, initiate water-related data sharing and use hydraulic, hydrological and economic modelling to improve the Kabul River Management Plan (Burki, 2013; Ahmadzai and McKinna, 2018). This idea was part of World Bank capacity-building projects in Afghanistan, such as the Afghanistan Water Resources Development Project (World Bank, 2018).
- *Hydropower memorandum of understanding.* In August 2013, the federal finance ministers of both countries signed an MoU to mutually develop a hydropower project on the Kunar river (SIWI, 2015; Malyar, 2016; Saeed *et al.*, 2016); however, this is yet to be implemented.
- *Data exchange.* The water and foreign affairs ministries of both Afghanistan and Pakistan were invited to Dubai by the World Bank in late 2014 to mutually develop techniques for exchanging water-related data and water allocation mechanisms in the Kabul river basin (Fig. 9.4). However, a

Fig. 9.4. Group photo of Afghan and Pakistan delegates at a workshop in Dubai, January 2020 (lead Pakistan).

follow-up meeting has not yet been called (SIWI, 2015).

- *Chinese support for joint hydropower development.* A trilateral meeting in 2015 was conducted between Afghanistan, China and Pakistan in Kabul, where China agreed to finance a joint hydropower project of 1500 MW on the Kabul river (SIWI, 2015). However, progress has been slow and despite multiple attempts, officials from both countries have failed to operationalize this initiative.
- *Flood control.* During a UNDP-hosted event in 2015 in Islamabad, MEW-Afghanistan and the Ministry of Climate Change in Pakistan agreed to collaborate on GLOF-related flood events (SIWI, 2015). Similarly, in June 2016, the Chief Executive of Afghanistan, Abdullah-Abdullah, expressed his interest to formalize water sharing in transboundary river basins with Afghanistan's neighbours so that all riparian countries could benefit (SIWI, 2015). During the event, the Chief Executive expressed a vision for Afghanistan to enhance its water storage capacity.

Indus Water Treaty. Even though the catchments of the Kabul, Kurram and Gomal basins are part of the greater Indus basin, the scope of the Indus Water Treaty of 1960 (IWT) does not include the rivers of these basins. The IWT governs the waters of the Indus river between Pakistan and India and has been famous for its stability during several armed conflicts between its party states (Dinar *et al.*, 2007). However, the IWT includes only those rivers which flow from India to Pakistan and does not include the Kabul river, which joins the Indus in the northwest of Islamabad (Chellaney, 2013). Since the IWT is exclusively concluded between Pakistan and India, it does not include Afghanistan and China as riparian basin states. Based on this reasoning, the IWT has been frequently criticized for laying the foundation for fragmented governance in the basin and is inconsistent with the principles of integrated water resources management (Zawahri and Michel, 2018).

Consequences of Non-co-operation on Shared Afghan–Pakistani Waters

Two primary adverse effects of non-co-operation. Although water co-operation has not yet materialized, cross-border engagement on water would deliver immediate value. Co-operation would address two present and recurring costs: (i) flood damage; and (ii) inefficient water management (Table 9.1). A flood early-warning system that communicates information on flood risk, for example, would provide value to both countries by enabling quicker adaptation to and reduced damage from floods. A second recurring cost in the status quo is the inability to optimally manage water due to the lack of provision of water data (e.g. on river flow). An ongoing exchange of water resources data and information would enable managers on both sides of the border to better plan and allocate water for enhanced benefits on an ongoing basis under present conditions.

Lack of co-operation brings future risk to Pakistan and Afghanistan. Although Pakistan is using a major portion of its water resources, the country may face the risk of losing access to water should Afghanistan plan and develop new dams and irrigation infrastructure (Hessami, 2017; Majidyar, 2018). The unilateral development of a shared river basin by an upper riparian (e.g. Afghanistan) has the potential to harm the quantity and quality of river flows to the downstream riparian (e.g. Pakistan).

Alteration in the freshwater flow to Pakistan could have adverse impacts on the livelihoods of much of the rural population (Azizi, 2007; Hanasz, 2011). Equally, however, expansion of water use infrastructure in Pakistan could establish a 'prior use' principle that forecloses future development opportunities. Unilateral development, if not undertaken co-operatively, could strain already tense relations between the two countries (Hanasz, 2011; Kakakhel, 2018). Issues related to equitable water sharing between Afghanistan and Pakistan are highly sensitive given the historical border dispute and political instability in the region (Favre and Kamal, 2004, p. 107; Hussain, 2011).

Barriers to Co-operation

Given the current costs and potential risks of non-co-operation, a valid line of inquiry can be directed to asking why there is no co-operation. Central to the lack of co-operation is the lack of trust between the two countries which is driven by the complex history between the two states. International relations between Afghanistan and Pakistan have been hampered by four specific issues that generate mistrust and which have led to this persistent absence of co-operation on water issues of the Kabul, Kurram and Gomal river basins: (i) boundary controversy; (ii) global and regional geopolitics; (iii) bundling water with other issues; and (iv) political instability.

Boundary controversy. In 1893, the Durand Line Agreement was signed between Afghanistan's King Abdul Rahman Khan and Sir Henry Mortimer Durand, the Foreign Secretary of British India. Since the partition of British India in 1947, Afghanistan and Pakistan have disagreed on the validity of the Durand Line as a formal international boundary between them. Afghanistan opposed the admission of Pakistan to the United Nations in 1947 (Sial, 2013) and claimed the areas east of Durand Line to the Indus river. As cited in various sources, Afghanistan puts forward several arguments against the Durand Line as a formal frontier. For example, it says the agreement was negotiated and concluded under duress (Mahmood, 2005) due to the nature of Afghan–British India relations at the time. Some have also argued that the Durand Agreement expired in 1993 and has lost its legality after its 100-year limit (Mahmood, 2005). Two well-known scholars of political history, Hassan Kakar and Azmat Hayat Khan, have even rejected the existence of any copy of the agreement executed by Abdul Rahman Khan (Kakar, 2006). Another point presented in most of the literature is that the map of the Durand Line has never been agreed or implemented. Pakistan rejects all these points and argues that the Durand Line Agreement of 1893 remains a valid international agreement which demarcates the international boundary between the two states.

Needless to say, both states have taken different positions on the validity of the Durand Line as an international border. For Afghanistan, acceptance of the Durand Line means the loss of the large Pashtun population and access to the sea. For Pakistan, the rejection of the Durand Line poses a serious threat to its territorial integrity. This longstanding conflict on the boundary has prevented friendly relations between the two states and has brought them to the brink of armed confrontation several times. Troublesome interstate relations have historically created mistrust and

Table 9.1. Costs and risks of non-co-operation.

Current costs	Future risks
• Failure to avoid or mitigate flood damage • Suboptimal water management	• Unilateral downstream (in Kabul river basin) development forecloses Afghanistan's rights as the upstream least-developed country. • Disruption to downstream livelihoods • Increased tensions between the two countries • Unilateral upstream development undermines downstream water management. • Asymmetric development will make the future settlement of water disputes less likely due to the existence of large agricultural and hydropower interests of the more developed country. • Unilateral developments may exacerbate the conflict between hydropower and irrigation or may constrain each other's potential.

blocked the way for any meaningful co-operation on other issues such as water.

Global and regional geopolitics. The partition of British India in 1947 and the emergence of Pakistan and India as independent states coincided with the beginning of the Cold War between the Soviet Union and the United States. Soon after its independence, Pakistan became an ally of the USA. In the early period of the Cold War, Afghanistan remained independent under King Zahir Shah, but later, in 1978, the pro-Soviet Communist parties overthrew the Republic of Duad and seized power. Pakistan was a US proxy and Afghanistan was a proxy for the USSR. Co-operation between pro-US and pro-Soviet governments in the Cold War was not extensive.

Global politics are evident in proxy conflicts. During the Cold War, both Afghanistan and Pakistan provided sanctuary or harboured and supported each other's political opponents. In the 1970s, Afghanistan provided shelter to the Baluch people who escaped from Pakistan's military operations in Baluchistan (Rubin and Siddique, 2006). Pakistan supported, armed and trained Afghan Mujahedin during the fight against the Soviets in the 1980s and the consequent civil war in Afghanistan. Pakistan continued to provide safe haven and support to the Taliban in the 1990s, 2000s and 2010s. The Soviet invasion of Afghanistan in 1979 tightened the US–Pakistan alliance even further. Therefore, both interstate and global geopolitics came into play in the already antagonistic and hostile relations between Afghanistan and Pakistan. Pakistan received substantial support from the USA, western Europe, Saudi Arabia and China for its proxy war against the Soviet Union in Afghanistan (Rubin and Siddique, 2006) and for nearly two decades (2002–2021), the Afghan government received major support from the USA.

Regional politics are manifested through the influence of the India–Pakistan rivalry. Strained relations between Afghanistan and Pakistan have continued in the post-Cold War era. They are now driven more by regional than by global geopolitics. This rivalry has led both countries to support allied factions. Pakistani politicians perceived the relatively expanded relations of the Afghan government with India as a threat of encirclement by India. In the post-9/11 era, Pakistan became America's ally in the 'war on terror' but it had maintained close ties with the Taliban to protect its security interests in Afghanistan against India. Ultimately, in the view of Pakistan, any Indian assistance to Afghanistan in terms of water resources management in the Kabul river basin poses another security threat which should be addressed in its bilateral relations with Afghanistan. A good example of this is Indian efforts to finance the construction of the Shatoot Dam in Kabul (Majidyar, 2018).

Bundling water with other issues. In the case of Afghanistan–Pakistan relations, the ruling elites in both states often conflate water with other interstate issues such as the Durand Line, India–Pakistan rivalry, transit, trade and security. Such an approach directly subsumes water under multiple foreign policy issues, causing water co-operation to stall until a long list of intractable issues are resolved. Examples of linking water with other issues include:

- In the post-9/11 era, when the Afghan government intended to commence the Shatoot Dam with financial support from India, to partially tackle the drinking water problem in Kabul City, the Pakistani elite and mass media perceived this as a threat from India (Ramachandran, 2018). However, internal Afghanistan discussions assumed that Pakistan was trying to link discussions on the dam to other issues (e.g. security, geopolitics and hydro-hegemony) between the two countries.
- The Afghan government has tried several times to speed up the Machalghu Dam construction on the Kurram river. However, Pakistan perceived this as a direct threat to its irrigation infrastructure and allegedly attempted to dismantle the project through its proxy insurgents (Ramachandran, 2018). The Kurram river irrigates about 80,000 acres of agricultural land in the Khuram Agency in Pakistan (Ali, 2016). The Kurram Tangi Dam on the Kurram river is already under construction in Pakistan. It is widely

believed in Afghanistan that Pakistan dismantled the Machalghu construction to protect its current and future uses of water from the Kurram river. In their view, it seemed that by dismantling such projects, Pakistan was pursuing a path towards hydro-hegemony, which would maintain both the status quo in water flow and gain geopolitical benefits.

Political instability. Frequent changes of government in Afghanistan and Pakistan also hinder the journey towards water co-operation. Any new government will need time to (i) gain international recognition and acceptance in the region and the world; and (ii) build internal consensus on the strategic benefits and importance of co-operation. In the current context in Afghanistan (from 15 August 2021 onwards), the Taliban government as a whole (or individual key politicians) can be expected to proceed with caution on water co-operation due to the fear of signalling tacit acceptance of the Durand Line. Ultimately, it may take years for the current government to embrace a nuanced foreign policy that allows it to strategically de-link water sharing issues from other issues with a common neighbour.

Potential benefits of co-operation

Despite the strained interstate relations between Afghanistan and Pakistan on water and other issues, transboundary water co-operation has always presented an opportunity for both countries to benefit. Six related but discrete benefits of co-operation can be discerned, which can help the two reluctant states initiate at least basic communication on water issues:

1. energy augmentation
2. flood risk reduction
3. more reliable flow for agriculture
4. climate change adaptation
5. peace and stability
6. macro-economic growth

Energy augmentation. The water resources of the Kabul river basin have tremendous hydropower potential, mainly in the Afghan portion. A World Bank (2018), for example, identified over 1100 MW of hydropower potential in the basin. The people in the basin and both states desperately need additional electricity (Yousaf, 2017). While Afghanistan could, in principle, unilaterally develop its water resources for national energy production, in-house financial capacities to support this development are unlikely to be sufficient. Afghanistan would thus require international financing, which would likely trigger the need for notification to downstream countries and basin co-operation. If undertaken co-operatively, hydropower infrastructure development could bring substantial additional electricity online for use in both countries (see Chapter 5).

Flood risk reduction. Floods cause immense damage to both countries, manifested in damage to property and loss of life. The Kabul river basin is among the most vulnerable regions to floods (Mayar *et al.*, 2020). For instance, the 2010 floods severely affected communities in both countries. These floods affected central and eastern parts of Afghanistan and several districts in Khyber Pakhtunkhwa Province (KPK), which are in the Kabul river basin (Yousaf, 2017). The 2010 floods caused about 2000 casualties and destroyed over 1.6 million houses in Peshawar, Charsadda, Mardan, Nowshera, Swabi and Swat districts of KPK in Pakistan (Iqbal *et al.*, 2018). There is obvious benefit to both countries in implementing a joint and co-operative flood early warning system, which could enable timely communications that allow for the avoidance of adverse flood impacts. Given the nature of river flows in the three basins, co-operation is critical to enable flood-inducing flow thresholds in upstream portions to be rapidly transmitted to downstream portions of the basins to enable preparation. A co-operative framework is critical to realizing these flood reduction benefits.

More reliable flows for agriculture. Lack of upstream water development (e.g. in the form of storage infrastructure) currently leaves both upstream and downstream agricultural water requirements dependent on natural flow variability. In some years, this means too little water is available while in other years there is too much. While seasonal peak flows occur each year, the Pakistani portions of the Kabul basin have experienced an increasing trend in both the frequency and magnitudes of floods. Upstream water development, if developed and operationalized co-operatively, could enable flow regulation that would provide more water supply assurance to Pakistan as the downstream neighbour. Equally, by storing water, it would increase water availability, presumably for agriculture in Afghanistan. Agriculture would gain from co-operative water resources development.

Climate change adaptation. By the end of the 21st century, the average temperatures in Kabul, Kurram and Gomal river basins are projected to increase by 2.6–5.1°C and the overall precipitation will increase by 4.5–12.2% (Khatiwada *et al.*, 2021). Increasing temperatures will significantly affect the snow and glacier melt and runoff to the Kabul river and its tributaries. It is estimated that 20% of the glaciers in the Kabul river basin will be depleted by the end of this century (Taraky, 2021). As a result of these changes, the availability of water for agriculture and domestic uses will be reduced, floods and droughts will become more frequent, and hydropower generation, biodiversity and sedimentation will also be negatively affected (Iqbal *et al.*, 2018; Khatiwada *et al.*, 2021). Co-operative basin management could be one of several tools to adapt to climate change. Integrated compilation of basin data, for example, and flexible allocation mechanisms can support the identification of water resource trends and optimal management responses.

Enhanced peace and security. Water co-operation of any form promotes peace and security at regional, interstate, national and provincial levels because co-operation provides a means of communication between stakeholders to discuss and address their interests, positions, risks, opportunities, challenges and solutions. Co-operation on water can reduce political tensions, improve security and help achieve foreign policy objectives (UNECE, 2015). In hydropolitical contexts such as Afghanistan and Pakistan, water co-operation can play a strategic confidence-building role, which can enhance regional co-operation and integration. In such a situation, Afghanistan can access seaports and Pakistan and India can gain access to Central Asian markets.

Regional economic growth. Co-operation in shared watercourses is widely acknowledged to produce benefits beyond the river (Sadoff and Grey, 2002). Chief among such benefits is a minimum platform for macroeconomic growth. Co-operation, and the stability that comes with it, provides a stable environment for much-needed investment, both international and national. Co-operation on water also has spillover effects in other sectors between co-operating countries which increases the overall pool of benefits derived from basins. Afghanistan and Pakistan, for example, may be willing to focus on their areas of comparative advantages to maximize production to support regional trade. Ultimately, the combined effects of these benefits are likely to manifest in an increase in regional economic growth.

Guiding Frameworks and Principles for Co-operation

Global and religious guidance concerning transboundary waters. The two most influential global conventions on transboundary waters are the 1992 UNECE Convention on the Protection and Use of Transboundary Watercourses and International Lakes, and the 1997 UN Convention on the Law of the Non-Navigational Uses of International Watercourses. The 1992 UNECE Water Convention came into force in 1996 and was subsequently converted into a global legal framework for enhancing transboundary water co-operation after the modification was adopted in February 2013. The 1997 UN Water Convention came into force in August 2014 and comprises the most widely referenced legal document on transboundary waters. In addition to global conventions, the importance and influence of customary law should be acknowledged. Finally, the relevance of religion in the Islamic world should not be underestimated. Islamic precepts should thus be considered insofar as they provide guidance on water sharing.

The 1992 UNECE Water Convention. The Water Convention aims to improve the protection and management of transboundary surface waters and groundwater. The main principles and instruments include:

- information exchange;
- notifying about planned measures and emergency situations;
- dispute resolution;
- obligation to co-operate;
- public access to data and information;
- equitable and reasonable utilization; and
- ecological principles (best available technology not entailing excessive costs, environmental impact assessment, ecosystem conservation and protection, the basin as the unit of management, prevention of pollution, monitoring, precautionary principle, polluter pays)

The 1997 UN Watercourses Convention. The main objective of the Convention is to safeguard the conservation, protection, management, development and use of transboundary water resources as well as to promote the sustainable and optimal use of shared water resources. Specifically, Articles 5 and 6 of the Convention on equitable and reasonable use and participation encourage the riparian states to develop and use transboundary water resources efficiently and equitably to achieve optimal and sustainable use and benefits by taking into account the interests of the riparian states. It also educates riparian states about the rights of use and the duty to co-operate on the protection and development of the watercourse. Criteria for the equitable distribution of water resources include geographic, hydraulic, climatic, economic and social (including attention to dependent populations), existing and potential water use, and its protection and use, among others.

Key guidance in the 1997 UN Watercourses Convention. The UN Watercourses Convention includes some important principles including: (i) political (information exchange, notification about planned measures and emergency situations, dispute resolution, and obligation to co-operate); (ii) social-relational (prior informed consent and equitable and reasonable use); and (iii) ecological principles (EIA, prevention of pollution, ecosystems conservation and protection, invasive species, and the basin as the unit of management). Each riparian state is obliged to provide data and information to the riparians of a shared watercourse (Article 9) concerning the condition and proposed uses (Article 12) so other states have sufficient time to study the intended water use and object if the use is considered harmful (Article 14). The Convention also calls upon states to take steps to prevent pollution-related damage (Article 21) and the introduction of invasive species (Article 22), and to enforce a responsibility on states that damage a shared watercourse to remedy or compensate (Article 22). The Watercourses Convention has specific provisions for damages to international watercourses due to natural calamities such as drought or erosion. It encourages participating states to inform riparians about the watercourse-related emergencies which might disturb them; for instance, water-borne diseases or floods (Article 27).

Customary water law. Customary international water law has been formally approved in the two UN water laws and codified in the 1966 informal Helsinki Rules and the 2002 Berlin Rules. Customary water law reflects existing state practice (Giordano and Wolf, 2003; Rahaman, 2009). It includes: (i) political principles (e.g. information exchange between riparians, obligation to co-operate, notifying riparian states about emergency situations and planned measures to avoid water flow interruption, limited territorial sovereignty, and dispute resolution); (ii) social-relational principles (e.g. equitable and reasonable use and capacity building); and (iii) ecological principles (e.g. protecting and preserving groundwater recharge zones and ecosystems) (McIntyre, 2011). Customary law imposes substantive and procedural obligations to co-operate in good faith (Shaarawy, 2016). Principles of customary law can potentially play a vital role in dispute resolution concerning water resource sharing among those countries not yet party to the 1997 Watercourses Convention (McIntyre, 2007).

Islamic law calls for co-operative water sharing. As expressly stated in the Quran, 'To Him belongs whatever is in the heavens and whatever is on the earth [including water]'

(Quran, 22:64) and in the following Hadith of Prophet Mohammad (peace be upon him), '... the people share three things: water, pasture and fire [energy]' (trans. Imame-Ahmad). In accordance with this fundamental Islamic rule, the waters of the Kabul river and its tributaries belong to all people living in both Afghanistan and Pakistan. To use this water for the benefit of the people in both countries, co-operation between the two states is required. Both Islamic and international water law consider water as a common resource which should be used co-operatively (Rezaee *et al.*, 2020). In Islamic law and jurisprudence, using water is generally viewed through the lens of a broad provision where every individual has the right to benefit from something that exists beyond private ownership (*mubah*). Furthermore, Islamic principles emphasize good conduct and conformity to established norms. Waste in any manner is prohibited and use cannot be detrimental to the resource or others as there is an order of priorities to be followed for water use, ranging from the basic right to quench your thirst, to commercial purposes. Other Islamic maxims – particularly those related to equitable use and ownership, avoiding harm, misuse of rights, ownership, the responsibility of consultation as well as the preservation of the environment and the ecosystem – all have implications for water planning and management.

Implications for the Afghan-Pakistani context. Since there is no existing transboundary water treaty between Afghanistan and Pakistan and neither of the two states are party to the UNWC or UNECE Convention of 1992, customary law can be applied. The equitable and reasonable principle or customary law provides relevant guidance for negotiations and a potential agreement between Afghanistan and Pakistan over shared water resources. In particular, the concept of limited territorial sovereignty upon which UNWC is based stipulates that riparian states have equitable rights to shared water use (Rieu-Clarke, 2005). Articles 5 and 6 also reflect the concept of limited territorial sovereignty.

Incentives and Catalysts for Water Co-operation

Earlier sections of this chapter have outlined how there are tremendous opportunities for joint benefit from Afghan–Pakistani co-operation on the Kabul, Kurram and Gomal rivers. Previous sections have also flagged a range of overarching geopolitical considerations that generally obstruct co-operation between Afghanistan and Pakistan and constrain progress on water. The aim of this section is to offer specific levers that can be used to trigger, catalyse, incentivize or motivate change to realize substantial potential benefits in the face of substantial contextual barriers. Four key levers of change are proposed:

1. unbundling water from other issues;
2. fostering buy-in to the value of basin management;
3. strong and sustained third-party facilitation;
4. urgency of climate change adaptation.

Unbundling water from other issues. The Indus basin possesses experience in unbundling water from other issues. Pakistan and India have experienced strained interstate relations throughout their existence which has prevented co-operation on most issues. Nonetheless, an Indus Water Treaty (IWT) was negotiated and concluded and peacefully divided the waters of the Indus between the two countries. The treaty has been largely successful for more than half a century, despite several armed conflicts and persistent geopolitical rivalry between the two parties. The main factors that motivated states to delink water from other issues included: (i) high dependence on water for irrigation requirements; (ii) appreciation of the shared nature of the resource; (iii) national development drives in both countries, which would benefit from assurance on water supply availability.

An important variable distinguishing Afghan–Pakistani water relations from Indo–Pakistani relations is the development trajectory of Afghanistan. Water development for irrigation in the colonial era was relatively high in India and Pakistan. On independence, both countries pursued ambitious national

Box 9.1. Learning from the Indus Water Treaty's experience in co-operating on water in the face of geopolitical challenges.

Negotiation, mediation, conclusion and implementation of the Indus Water Treaty (IWT) offer a unique case because it was negotiated and entered into during a period of high interstate tensions, poor relations, active conflicts, and acute mistrust between India and Pakistan. The IWT has remained stable amid several armed conflicts and continuing geopolitical rivalry. The World Bank, as a third party to the dispute, played a tremendously significant mediatory and financial role. However, it does not mean that the disputing parties' roles and involvement were not important.

There are similarities between the current Afghanistan–Pakistan interstate relations and the hostile India–Pakistan relations of that time. However, there are some key differences between the two and several lessons that Afghanistan and Pakistan could learn while settling Kabul river basin issues.

Separation of water from the broader interstate political issues would help both states to engage in discussions on the Kabul river. Discussions based on technical data and information rather than politics helped the two riparian states of India and Pakistan agree to facilitation from the World Bank (Wolf and Newton, 2013).

Begin with direct negotiations if possible. Initially, Pakistan and India began with direct negotiations, which showed their political will for resolution of the dispute. After a deadlock occurred in their interactions, the World Bank entered the scene.

Share water rather than divide it. One lesson that Afghanistan and Pakistan should *not* learn from the IWT is the separate management of water. IWT has not laid out a legal and institutional basis for joint and co-operative management but it has instead created an independent control mechanism for each riparian over certain tributaries of the Indus river. The partition of the Indus originated in the pre-partition era and was escalated by the partition of British India in 1947. Therefore, there are some differences with the hydro-geographical settings of the Kabul river and its tributaries.

Secure third-party mediation. Another lesson to be learned from the IWT is the necessity of involving a third party to play mediatory and financial roles to facilitate negotiations and fund a development plan for the basin. The World Bank as a third party had significant leverage over the parties and played an incredible role in keeping the parties together and providing a pragmatic solution which was mutually acceptable (Salman, 2013). A similar third party that can play both mediatory and financial roles should be considered if Afghanistan's and Pakistan's negotiations on the Kabul river do not lead to a fair, sustainable and equitable consensus. However, it is important to consider the timing of third-party involvement.

development objectives. Afghanistan, by contrast, did not experience the same level of development in the colonial era and was not able to realize extensive national development progress in its period of independence. These dynamics result in a situation where co-operation on water may not constitute an urgent need on Afghan–Pakistani tributaries that it did on Indo–Pakistani portions of the Indus basin. Nonetheless, ample opportunity remains for energy production through water development, which countries desperately need. It may be fruitful to explore ways to unbundle water from other issues as a means to secure these benefits in the short term while countries can continue to resolve broader disagreements in the long term. The nearby Indus Water Treaty exemplifies hydro-diplomacy amid geopolitical challenges (Box 9.1).

Basin management buy-in. Afghan–Pakistani co-operation on the three Indus tributaries is predicated on the foundation that the two countries appreciate the value of basin management. This appreciation should not be assumed. In the case of the IWT, a delegate from the Tennessee Valley Authority (TVA) – a successful basin management organization in the American South – visited the region to communicate the importance of basin management. While specific benefits that can be obtained from co-operative management of Indus tributaries can certainly be presented to the two countries (as elaborated in potential benefits of co-operation), direct presentation of benefits is often insufficient to motivate change in shared waters (Subramanian *et al.*, 2012). Securing deeper buy-in to the concept of basin management through, for example, joint study tours to effective river basin organizations such as the TVA and the Murray Darling Authority, or capacity development activities on international experiences in shared water management, can help countries see and appreciate the value of managing basins as a unit. Successful river basin organizations seem to have learned some of the following lessons.

Strong and sustained third-party facilitation. The role of third parties in negotiations is widely recognized. The IWT process highlights the necessity of involving a third party to mediate and broker negotiations for a shared watercourse. In the case of the IWT, the World Bank had significant leverage over the parties and played a critical role in keeping the parties together and providing a pragmatic and mutually acceptable solution (Salman, 2013). World Bank involvement in shared water issues often brings mutual incentives such as: (i) a stable institutional foundation enables the World Bank to lend funds for infrastructure; and (ii) countries can secure benefits from infrastructure. In the case of Afghan–Pakistani shared waters, an open question is whether the World Bank will see the potential for lending given that most infrastructure sites are in Afghanistan where there has not been a stable track record. Other third parties, with either different incentives or a higher threshold for risk, may nonetheless show an interest in the region and serve as a third-party facilitator. Western governments, for example, may see value in promoting regional peace and security and transboundary water co-operation may be one pathway towards that goal. Whatever the case, given the gestation period of transboundary water interactions, third-party facilitation would need to be strong and sustained.

The urgency of climate change adaptation. A final co-operation driver may be climate change adaptation. Long-term effects of climate change may dwarf the current severe challenges. Climate change may provide a sense of urgency to work together that can be harnessed to promote co-operation. Climate change also brings a suite of opportunities for financial support that can help incentivize and sustain co-operation. Ultimately, it may be wise to employ more than one of the change levers described here in a co-ordinated way to unblock co-operation.

Potential Modalities of Benefit-sharing

Stages of water co-operation. Sadoff and Grey (2007) divided co-operation into three stages (Table 9.2), which largely follow those outlined by Boisson *et al.* (2013). The first stage captures mechanisms for which implementation requires interstate *co-ordination*, but relatively few institutional efforts. For example, jointly gathering and exchanging data and information require co-ordination (Sadoff and Grey, 2007). The second stage captures *collaboration*, which results when national plans are adjusted either to secure gains or to mitigate harm in another riparian country (ibid.). Diverting upstream water may, for example, effectively alleviate flood risks downstream. The third stage reflects mechanisms that require effective *joint action* and riparians acting as partners in the design, investment and implementation of transboundary river development (ibid.). This level of co-operation is often formalized by a joint body with a mandate.

Scaling up water co-operation. Mirumachi (2007) and Zeitoun and Mirumachi (2008) often speak of interactions that portray both a gradational co-operation spectrum on one axis and a conflict spectrum on the other. The co-operation spectrum advances from *ad hoc* co-operation to co-operation on technical issues, then proceeds to risk-averse co-operation and to risk-taking co-operation. Their co-operative steps effectively incorporate the element of risk, which has been identified as a key factor in state decision-making related to engagement on transboundary waters (Subramanian *et al.*, 2012).

Co-operation on Afghan–Pakistani Indus tributaries is starting from zero. As stated above, there is currently no institutionalized water interaction between Afghanistan and Pakistan. Water management on Indus tributaries is unilateral and *ad hoc*. In this context of 'no co-operation', a key consideration to phasing in water co-operation is how this co-operation is sequenced. Careful consideration of sequencing allows riparians to select focal points for co-operation, engage in a consensus-based bargaining process, and move from one equilibrium to another. Sequencing also supports the process of international co-operation by breaking down decision-making into small steps (Young, 2010, p. 1).

Co-operative first steps. While there are undoubtedly more benefits to be achieved through more intense co-operation focused on joint actions like infrastructure development, the degree to which Afghan–Pakistani shared basins are ready for this is debatable. Instead, it may make sense to start small to generate trust on issues in which stakes, time and energy investments are lower. Such efforts in the initial period of co-operation would include data and information exchange, and flood early warning. Specific initial steps could focus on confidence-building measures to kickstart a programme that ultimately realizes these objectives, such as:

- A *joint study* on issues explicitly or implicitly tied to water resources undertaken by the two governments. The central focus of the study should be treated as a topic to motivate collaboration rather than polarization. As such, allocation should be avoided. Climate change, by contrast, could feature prominently. This study could build on university partnerships.
- *Study tours* can be advanced to foster country interaction and trust. Locations with effective basin management can be selected as a focus so that participants gain an appreciation for the benefits that can be derived from effective basin management.
- *Capacity building in negotiation skills* can be provided through a joint workshop or series of joint workshops. Among the focus of such a programme could be examples in which countries have advanced their national interests by unbundling foreign policy issues.

Focus on specific issues in specific basins. Another way to foster small co-operative steps is to drill down to specific issues in specific basins. While there are many common issues and challenges across the Kabul, Kurram and Gomal, there are also differences. The Kabul is the largest and most populous basin and it may also be the most contentious. Afghan and Pakistani governments may therefore view Kurram and Gomal as less strategic and thus less risky to test co-operation with the other side. In the Gomal and Kurram basins, for example, the two countries could conduct a joint study on the effects of climate change on floods. This study can be focused on how floods became more severe and frequent compared to the past, due to the changing climate. An even smaller co-operative step in both Kurram and Gomal could be a joint study of floods only.

Partnership opportunities for financial, technical and capacity support. To advance co-operation, it will be necessary to mobilize third-party support. Potential third parties include some of those referred previously (e.g. World Bank, USAID), who have already made efforts to advance co-operation. Other actors could include organizations like Green Cross International or GIZ. Whichever actors engage, they must strive for long-term commitment rather than short-term activities. One partner model worth considering is the Nile Basin Initiative. The World Bank established a multi-donor trust fund into which several

Table 9.2. Levels of co-operation on shared waters.

Source	No co-operation	→		High co-operation
Sadoff and Grey (2007)	Unilateral	Co-ordination	Collaboration	Joint activity
Mirumachi (2007); Zeitoun and Mirumachi (2008)	Ad hoc	Technical	Risk-averting	Risk taking
Synthesized approach	No co-operation	Co-ordination and assistance; often technical	(Active) collaboration	Joint action

donors can contribute. This model has enabled over two decades of funding to Nile basin co-operation. While this may not reflect well on the ability of Nile-focused co-operation to sustain itself, the continuity of funding is in line with the long gestation period and commitment often required for transboundary water engagement.

References

Abderrahman, W.A. (2000) Application of Islamic legal principles for advanced water management. *Water International* 25(4), 513–518. DOI: 10.1080/02508060008686865.

Ahmad, S. (2010) Towards Kabul Water Treaty: managing shared water resources: policy issues and options. IUCN, Karachi, Pakistan.

Ahmadzai, S. and McKinna, A. (2018) Afghanistan electrical energy and trans-boundary water systems analyses: Challenges and opportunities. *Energy Reports* 4, 435–469. DOI: 10.1016/j.egyr.2018.06.003.

ALEP (2011) *Afghanistan Legal Education Project (ALEP)*. Available at: https://www-cdn.law.stanford.edu/wp-content/uploads/2016/03/ALEP-Law-of-Afghanistan-3d-Ed_English.pdf (accessed July 2022).

Ali, F. (2016) Pakistani officials call for water agreement with Afghanistan. *The Third Pole*. Available at: https://www.thethirdpole.net/2016/09/21/pakistani-officials-call-for-water-agreement-with-afghanistan/ (accessed 28 September 2022).

Atef, S.S., Sadeqinazhad, F., Farjaad, F. and Amatya, D.M. (2019) Water conflict management and cooperation between Afghanistan and Pakistan. *Journal of Hydrology* 570, 875–892. DOI: 10.1016/j.jhydrol.2018.12.075.

Azam, S. (2015) Kabul River Treaty: a necessity for peace-n-security between Afghanistan and Pakistan, and peace in South Asia. *Gomal University Journal of Research* 31(2).

Azizi, P.M. (2007) Special lecture on water resources in Afghanistan.

Biswas, A. (2013) Durand Line: history, legality and future. Occasional paper. Available at: https://www.files.ethz.ch/isn/170887/Durand%20Line_History%20Legality%20%20Future_Final.pdf (accessed 21 July 2022).

Boisson, de L., Leb, C. and Tignino, M. (2013) *International Law and Freshwater*. Edward Elgar, Northampton, Massachetts.

Burki, S.J. (2013) Economics of Pak-Afghan relations. ISAS Working Paper No. 176, Institute of South Asian Studies. National University of Singapore.

Chellaney, B. (2013) *Water, Peace and War: Confronting the Global Water Crisis*. Rowman & Littlefield, London.

Dinar, A., Dinar, S., McCaffrey, S. and McKinney, D. (2007) Bridges over water understanding transboundary water conflict, negotiation and cooperation. *World Scientific Series on Energy and Resources Economics* 3.

Favre, R. and Kamal, G.M. (2004) Watershed atlas of Afghanistan: working document for planners, parts I and II. Government of Afghanistan, Ministry of Irrigation, Kabul.

Giordano, M.A. and Wolf, A.T. (2003) Sharing waters: post-Rio international water management. In: *Natural Resources Forum*, Vol. 27(2). Blackwell Publishing Ltd, pp. 163–171.

Hanasz, P. (2011) The politics of water security in the Kabul river basin. *Future Directions International*. Available at: http://www.cawater-info.net/afghanistan/pdf/hanasz.pdf (accessed 21 July 2022).

Hessami, E.B. (2017) Afghanistan's water plans complicated by worried neighbors. *New Security Beat* [blog of the Environmental Change and Security Program]. Available at: https://www.newsecuritybeat.org/2017/03/afghanistans-water-plans-complicated-worried-neighbors/ http://www.cawater-info.net/afghanistan/pdf/hanasz.pdf (accessed 21 July 2022).

Hessami, E.B. (2018) As Afghanistan's water crisis escalates, more effective water governance could bolster regional stability. *New Security Beat* [blog of the Environmental Change and Security Program]. Available at: https://www.newsecuritybeat.org/2018/07/afghanistans-water-crisis-escalates-effective-water-governance-bolster-regional-stability/ (accessed 10 April 2022).

Hussain, Z. (2011) Sources of tension in Afghanistan and Pakistan: a regional perspective. *CIDOB Policy Research Project*. Available at: https://www.cidob.org/en/publications/publication_series/project_

papers/stap_rp/policy_research_papers/sources_of_tension_in_afghanistan_and_pakistan_a_regional_perspective (accessed 21 July 2021).
Iqbal, M., Dahri, Z., Querner, E., Khan, A. and Hofstra, N. (2018) Impact of climate change on flood frequency and intensity in the Kabul River Basin. *Geosciences* 8(4), 114. DOI: 10.3390/geosciences8040114.
Janjua, M.Q. (2009) In the shadow of the Durand Line: security, stability, and the future of Pakistan and Afghanistan. Available at: https://apps.dtic.mil/dtic/tr/fulltext/u2/a501684.pdf (accessed 21 July 2022).
Kakakhel, S. (2017) Afghanistan-Pakistan Treaty on the Kabul River Basin? *The Third Pole*. Available at: https://www.thethirdpole.net/en/2017/03/02/afghanistan-pakistan-treaty-on-the-kabul-river-basin/ (accessed 21 July 2022).
Kakakhel, S. (2018) Agreements and institutions related to shared rivers within South Asia and beyond. In: Ahmed, I. (ed.) *South Asian Rivers.* The Anthropocene: Politik-Economics-Society-Science Series, Vol. 21. Springer, Cham, Switzerland. Available at: https://doi.org/10.1007/978-3-319-67374-5_3 (accessed 28 September 2022).
Kakar, M.H. (2006) *A Political and Diplomatic History of Afghanistan, 1863–1901*. Brill, Leiden, The Netherlands.
Kaura, V. (2017) *The Durand Line: A British Legacy Plaguing Afghan-Pakistani Relations*. Middle East Institute. Available at: https://www.mei.edu/publications/durand-line-british-legacy-plaguing-afghan-pakistani-relations (accessed 21 July 2022).
Khan, H.F., Yang, Y.C.E. and Wi, S. (2020) Case study on hydropolitics in Afghanistan and Pakistan: energy and water impacts of Kunar River development. *Journal of Water Resources Planning and Management* 146(9), 05020015. DOI: 10.1061/(ASCE)WR.1943-5452.0001262.
Khatiwada, K., Pradhananga, S. and Nepal, S. (2021) Potential implications of future climate change and extremes on health, agriculture, water, and cryosphere sectors in the Kabul River Basin. DOI: 10.21203/rs.3.rs-645615/v1.
Mahmood, T. (2005) The Durand Line: South Asia's next trouble spot. Master's thesis, Naval Postgraduate School, Monterey, California.
Majidyar, W. (2018) Afghanistan and Pakistan's looming water conflict. *The Diplomat*. Available at: https://thediplomat.com/2018/12/afghanistan-and-pakistans-looming-water-conflict/ (accessed 21 July 2022).
Malik, T. (2019) Pak-Afghan water issue: a case for benefit-sharing. *Policy Perspectives* 16(1), 77–98. DOI: 10.13169/polipers.16.1.0077.
Malyar, I. (2016) Transboundary water institutions in developing countries: a case study in Afghanistan. MSc thesis, Oregon State University.
Mayar, M.A., Asady, H. and Nelson, J. (2020) River flow analyses for flood projection in the Kabul River Basin. *Central Asian Journal of Water Research (CAJWR)* Центральноазиатский Журнал Исследований Водных Ресурсов 6(1), 1–17.
McCarthy, J. and Mustafa, D. (2014) Despite the best intentions? experiences with water resource management in northern Afghanistan. *Academia*. Available at: https://www.academia.edu/70014266/Despite_the_best_intentions_Experiences_with_water_resource_management_in_northern_Afghanistan (accessed 21 July 2022).
McIntyre, A. (2007) *Participatory Action Research*. Sage Publications, London.
McIntyre, O. (2011) The World Court's ongoing contribution to international water law: The Pulp Mills Case between Argentina and Uruguay. *Water Alternatives* 4(2), 124–144.
Mirumachi, N. (2007) Fluxing relations in water history: conceptualizing the range of relations in transboundary river basin. In: *Pasts and Futures of Water: Proceedings from the 5th International Water History Association Conference*, Tampere, Finland, June 13–17, 2006.
Omrani, B. (2009) The Durand Line: history and problems of the Afghan-Pakistan border. *Asian Affairs* 40(2), 177–195. DOI: 10.1080/03068370902871508.
Panikkar, B., Zia, A., Sgorbati, S., Cohen, M., Abid, M. *et al.* (2019) Transboundary water governance in the Kabul River Basin: implementing environmental and public diplomacy between Pakistan and Afghanistan. *Complexity, Governance & Networks* 5(1), 101. DOI: 10.20377/cgn-80.
Pervaz, I. and Khan, S. (2014) Brewing conflict over Kabul River: policy options for legal framework. Available at: https://ndu.edu.pk/issra/issra_pub/articles/issra-paper/ISSRA_Papers_Vol6_IssueII_2014/03-Brewing-Conflict-over-Kabul-River.pdf (accessed 21 July 2022).
Qureshi, A.S. (2002) *Water Resources Management in Afghanistan: The Issues and Options*. International Water Management Institute, Colombo, Sri Lanka.

Rahaman, M.M. (2009) Principles of transboundary water resources management and Ganges treaties: an analysis. *International Journal of Water Resources Development* 25(1), 159–173. DOI: 10.1080/07900620802517574.

Ramachandran, S. (2018) India's controversial Afghanistan dams. *The Diplomat*. Available at: https://thediplomat.com/2018/08/indias-controversial-afghanistan-dams/ (accessed 21 July 2022).

Rezaee, M., Hearns, G. and Leb, C. (2020) Managing Blue Gold: a comparative study of Islamic Law and International Water Law: 27 Water Law. Available at: http://www.internationalwatersgovernance.com/uploads/1/3/5/2/13524076/2020_rezaee_et_al_islamic___int_wat_law.pdf

Rieu-Clarke, A. (2005) *International Law and Sustainable Development*. IWA Publishing.

Rubin, B. and Siddique, A. (2006) *Resolving the Pakistan-Afghanistan Stalemate*. United States Institute for Peace (USIP), Washington, DC.

Sadoff, C. and Grey, D. (2002) Beyond the river: the benefits of cooperation on international rivers. *Water Policy* 4, 389–403. DOI: 10.1016/S1366-7017(02)00035-1.

Sadoff, C.W. and Grey, D. (2007) Cooperation on international rivers: a continuum for securing and sharing benefits. *Water and Energy Abstracts* 16(3), 25–26.

Saeed, B.A., Hassani, S. and Malyar, I. (2016) *Transboundary Basin Management under Conditions of Latent Conflict: A Multi-Sectoral and Multi-Disciplinary Approach towards the Kabul River Basin*. Heinrich Boll Stiftung, Berlin.

Salman, S.M. (2013) Mediation of international water disputes – the Indus, the Jordan, and the Nile Basins interventions. In: Boisson, de L., Leb, C. and Tignino, M. (eds) *International Law and Freshwater*. Edward Elgar, Cheltenham, UK.

Shaarawy, S. (2016) The role of customary international water law in settling water disputes by mediation: an examination of the Indus river and Renaissance dam disputes.

Sial, S. (2013) Pakistan's role and strategic priorities in Afghanistan since 1980. Norwegian Peacebuilding Resource Center (NOREF). Available at: https://www.files.ethz.ch/isn/165432/9bc5b02e91c5a9b8ba49a5c46dbfd41a.pdf (accessed 21 July 2022).

SIWI (2015) MiCT_SIWI_Orphan-River_Final.pdf. Available at: https://mict-international.org/wp-content/uploads/2016/01/MiCT_SIWI_Orphan-River_Final.pdf

Subramanian, A., Brown, B. and Wolf, A. (2012) *Reaching Across the Waters: Facing the Risks of Cooperation in International Waters*. World Bank Group Number 13078.

Taraky, Y. (2021) Transboundary water resources in the context of developmental goals and climate change: the case of the Kabul River Basin. Doctoral dissertation, University of Guelph, Canada.

Tariq, M.A.U.R., van de Giesen, N., Janjua, S., Shahid, M.L.U.R. and Farooq, R. (2020) An engineering perspective of water sharing issues in Pakistan. *Water* 12(2), 477. DOI: 10.3390/w12020477.

UN (1997) Convention on non-navigational uses of international watercourses. Available at: https://legal.un.org/ilc/texts/instruments/english/conventions/8_3_1997.pdf

UNECE (2015) *Policy Guidance Note on the Benefits of Transboundary Water Cooperation:Identification, Assessment and Communication*. Convention on the Protection and Use of Transboundary Watercourses and International Lakes, United Nations Economic Commission for Europe.

UNECE (2016) *The Global Opening of the 1992 Water Convention*. ECE/MP.WAT/43/Rev.1. Available at: https://www.unece.org/fileadmin/DAM/env/water/publications/WAT_The_global_opening_of_the_1992_UNECE_Water_Convention/ECE_MP.WAT_43_Rev1_ENGLISH_WEB.pdf (accessed 28 September 2022).

Vick, M.J. (2014) Steps towards an Afghanistan–Pakistan water-sharing agreement. *International Journal of Water Resources Development* 30(2), 224–229. DOI: 10.1080/07900627.2014.886471.

Weinthal, E., Troell, J.J. and Nakayama, M. (2014) *Water and Post-Conflict Peace-Building*. Earthscan. DOI: 10.4324/9781849775809.

Wolf, A.T. and Newton, J.T. (2013) The Indus Treaty: conflicting riparians and the role of a third party. In: Earle, A., Jagerskog, A. and Ojendal, J. (eds) *Transboundary Water Management: Principles and Practice*. Earthscan, pp. 195–197.

World Bank (2018) *Afghanistan TO 2030 Priorities for Economic Development Under Fragility*. Available at: http://documents.worldbank.org/curated/en/156881533220723730/pdf/129161-WP-P157288-Afghanistan-to-2030-PUBLIC.pdf (accessed 28 September 2022).

Yoffe, S., Wolf, A. and Giordano, M. (2003) Conflict and Cooperation over international waters: indicators of basins at risk. *Journal of American Water Resources Association*. Available at: https://onlinelibrary.wiley.com/doi/abs/10.1111/j.1752-1688.2003.tb03696.x

Young, I.M. (2010) Responsibility and global labor justice. In: Ognjenovic, G. (ed.) *Responsibility in Context*. Springer, Dordrecht, pp. 53–76.

Yousaf, S. (2017) Kabul River and Pak-Afghan relations. *Central Asia Journal* 80, 102–103. Available at: http://journals.uop.edu.pk/papers/06%20Suliman%20Yousaf.pdf (accessed 21 July 2022).

Yousafzai, I.A. and Yaqubi, H. (2017) The Durand Line: its historical, legal and political status. Available at: http://pu.edu.pk/images/journal/history/PDF-FILES/6_54_1_17.pdf (accessed 21 July 2022).

Zawahri, N. and Michel, D. (2018) Assessing the Indus Waters Treaty from a comparative perspective. *Water International* 43(5), 696–712. DOI: 10.1080/02508060.2018.1498994.

Zeitoun, M. and Mirumachi, N. (2008) Transboundary Interactions I: reconsidering conflict and cooperation. *International Environmental Agreements* 8, 297–316.

10 Conclusions and Recommendations

Jonathan Lautze, Asadullah Meelad and Muhammad Azeem Ali Shah

Chapter Overview

This chapter reviews some of the book's overarching themes and main findings. Overarching themes include the recognition of persistent hardship and inequality, particularly in the Afghan portions of the basins and substantial knowledge gaps that need to be filled to address existing challenges. Key threads relate to trends in demography, climate, land and water use, and institutions that may accentuate challenges. The chapter concludes that more than incremental change will be needed to support the progress necessary to provide a basic level of human development. Key to such progress is transboundary co-operation between Afghanistan and Pakistan and re-engagement with the international community to create a context that attracts investment that catalyses desperately needed economic progress.

Overarching Themes

Afghan–Pakistani shared waters, particularly the Kabul, Kurram and Gomal rivers, carry historic, current and future significance. The waters of the Kabul, Kurram and Gomal rivers have long been central to domestic water supplies, food security and livelihoods for the people who live here. These shared rivers have, at the same time, driven chronic instability through the recurrent impacts of floods and droughts. Currently, these shared basins are home to approximately 43 million people and provide more than 26 bcm of water per year, on average. The populations, area and water resources of these basins, in practice if not principle, are shared between Afghanistan and Pakistan. Internationalization of these basins has served to increase their importance, as their management now carries regional geopolitical implications. The importance of these shared rivers will further increase due to the increasing impacts of climate change, rapid demographic growth and increasing demand for freshwater resources.

Review of the human aspects highlights hardship and inequality The recent regime change in Afghanistan and political changes in Pakistan have exacerbated human development conditions. The figures presented in this book may not fully reflect latest developments. However, finding accurate information after the recent changes is difficult. As outlined in Chapter 3, annual per capita income in the three basins may be consistent with national averages in the two countries (approximately US$500 in Afghanistan and US$1200 in Pakistan). In contrast, levels of poverty are presumed to be extremely high, as much as 50% of the population in the Afghan portions of the three basins and of the order of 20% in the Pakistani portions. In light of these realities, levels of human development as measured by the Human Development Index remain consistent with those of the bottom quarter of the world's nations. While human hardship is real and pervasive throughout the basins, conditions are consistently the most severe in the Afghan portions.

DOI: 10.1079/9781800622371.0010

There is abundant opportunity. There appears to be ample untapped potential to harness additional benefits in the three basins. As described in Chapter 5, the Kabul basin alone generates nearly 16 bcm of water on average per year, making it the largest tributary to the Indus river. But the Kabul, Kurram and Gomal basins have inadequate water storage infrastructure (e.g. large and small reservoirs). In the Kabul basin, for example, there are only three active water storage facilities (Sarobi, Naghlu and Warsak) that collectively impound only one bcm of water. This equates to less than 10% of the average annual runoff. While the transboundary nature of the three basins invariably complicates efforts to develop, the severity and pervasiveness of human hardship heightens the need to find ways to realize co-operative development.

Knowledge gaps are substantial. Improving conditions in the three basins will depend on formulating solutions that respond to specific contexts based on an inclusive collection and synthesis of data and information. However, as acknowledged in virtually every chapter, major knowledge gaps remain. Chapter 4 emphasized the high level of uncertainty associated with predicting climate change impacts, for example, and Chapter 6 identified a range of data gaps associated with specific groundwater parameters such as renewable yield, level of abstraction and groundwater level trends. Related to obvious data gaps is the paucity of data integration at a basin level needed to generate an inclusive knowledge foundation for the basin. If, as appears likely, water resources development needs to be undertaken through a basin lens, it will be essential to characterize the status of the basin to identify changes that might follow from such developments.

This is the first book dedicated to these basins and the first to characterize conditions in the three basins of the Kabul, Kurram and Gomal rivers. As outlined in Chapter 2, this is also the first book to pursue a co-production approach that involved regional experts from Afghanistan and Pakistan as well as international experts based around the world. Extensive efforts were made to obtain and compile data from international databases, national authorities and academic bodies in the two countries, as well as through personal networks of the authoring and editorial teams. Currently, this book provides a snapshot of the state of the Kabul, Kurram and Gomal river basins which, hopefully, will support the pursuit of solutions for improving basin management in ways that advance human wellbeing.

Main Findings

This book identifies challenges and opportunities in the shared Kabul, Kurram and Gomal systems. In their chapters, the authors have made every possible effort to consider the multidimensional nature of basin conditions, from socio-economics and demography to climate variability and change, surface and groundwater use and management, and institutions and governance from both a national and transboundary perspective. This is believed to be the first book that covers the whole basins – reflecting the inclusion of both Afghan and Pakistani portions. Ultimately, while the book illustrates that the challenges are substantial, equally evident is that ample opportunities exist to improve conditions.

Knowledge is a key enabler. While the elaboration of specific collaborative actions and solutions is the central focus of future work, perhaps undertaken as part of Track II diplomacy between the two countries, the information in this book provides clues that point to key areas that can facilitate improvements in water management within these shared systems and maximize the benefits derived from those improvements. Central to the elaboration of opportunities is improving the current knowledge base by developing data-collection infrastructure, expanding the depth of data and integrating data from both countries to enable a clearer understanding of current trends and potential. Trends that require further investigation include issues like changes in mean flows on both sides of the border and groundwater levels. Potential actions that merit additional investigation include the study of the impacts of climate change, options for energy sharing and transboundary flood early warning.

Main findings from each chapter. While it is always tempting to point to additional data and information that should be obtained and processed to more fully characterize conditions and formulate solutions, existing information does provide insights, even if incomplete. Major findings have been reported in Chapters 3 through 9, which speak to key aspects of resources management in the three basins including socio-economics, surface and groundwater availability and use, and national and transboundary institutions and governance.

Socio-economics and demography. Chapter 3 outlines conditions in each basin. The three river basins are largely rural but the population density is not insignificant. In Afghanistan, the population density is 22 persons per km^2 in the Gomal basin, 111 in the Kabul basin and 127 in the Kurram basin. In Pakistan, there are 146 inhabitants per km^2 in the Gomal basin, 254 in the Kabul basin and 327 in the Kurram basin. The higher population in the Pakistani portions typically earn higher incomes but people living on both sides face severe conditions including high poverty, high vulnerability to climate change, and some level of food insecurity.

Climate. Chapter 4 outlines the evolving climate realities in the basins. Two particularly concerning trends are noted. First, decreasing snowfall due to increasing temperatures is likely to result in a more variable river flow regime which will, in turn, require more regulation and management to enable a stable water supply. Second, increasing rainfall due to climate change is likely to increase the flood risk in the basin as peak flow may increase in volume and frequency. While such trends no doubt vary spatially in and across the three basins, it is clear that climatic change poses an increased risk to populations that are vulnerable due to poverty, food insecurity and other challenges, as outlined in Chapter 3.

Surface water hydrology. Chapter 5 outlines conditions related to surface water. One important issue identified is the sparseness of flow-gauge stations, particularly on the Afghan sides of the three basins. Other concerns are growing water quality challenges across several key parameters. The most significant message, however, is the dearth of water storage infrastructure which results in immense lost opportunity for energy production, and an appreciable increase in flood risk due to very limited ability to control river flows. There are minor variations across the three basins but, given its size, the Kabul basin would appear to present the greatest potential for water infrastructure development.

Groundwater. Chapter 6 outlines groundwater conditions. Limitations on groundwater data are extreme, particularly in the Afghan portions of the three basins. This is a concern because enhanced groundwater monitoring could play an important role in the better management of this resource. Nonetheless, the available data indicate some concerning trends. Pakistan withdraws groundwater at an unsustainable rate. In the Kabul basin, specifically in the upper and central portions of the basin, groundwater is being depleted at a faster rate than in the lower portions of the basin. A major catalyst for this excessive withdrawal is the insufficient application of rules and regulations for groundwater extraction. The situation is most likely similar in the Kurram and Gomal basins.

Land and water use. Chapter 7 outlines realities related to land and water use in the basins. The available data on water depletion relative to renewable water resources suggests substantial volumes of unused water, particularly in the Kabul basin. Across the three basins, agriculture is the dominant land use, totalling some 300,000 ha in the Afghan portions and at least 1,000,000 ha in the Pakistani portions. While precise volumes of water depletion associated with agriculture could not be determined, one assumes these are the largest productive uses of water. Also notable is the increase in urban water use, particularly in Afghanistan. Cities such as Kabul and Peshawar will require growing consideration in water management planning.

Institutions and governance. Chapter 8 reviews laws and policies relevant to water in the two countries. The authors illustrate how institutional frameworks related to water in the two countries are at something of a crossroads. Pakistan has made considerable progress that culminated in the 2020 Khyber Pakhtunkhwa Water Act and the creation of a KPK Water Commission. A more comprehensive 2018 National Water Policy is still awaiting implementation and the KPK Water Act remains to be incorporated into existing governance structures. All the existing frameworks are vague on transboundary water sharing. On the Afghan side, some progress was made under the 2002–2021 regime, such as the 2009 and 2020 Afghan Water Laws. However, the degree to which formal laws influence on-the-ground realities in rural areas of the Kabul, Kurram and Gomal basins is open to question. The recent (2021) regime change is raising new questions about the continued applicability of such progress.

Water co-operation. Chapter 9 focuses on the potential for co-operation in the shared Kabul, Kurram and Gomal basins. The authors found there is currently little co-operation on water between Afghanistan and Pakistan, and the major barriers to water co-operation are only tenuously related to water; for example, border controversies and geopolitical issues. There are, nonetheless, clear issues around which co-operation could bring direct benefit, such as flood mitigation now and hydropower in the future. Several incentives, if well framed, could be used to advance co-operation. One incentive could be derived from Islamic law, whose principles support sharing water and other resources. Given the scope of the challenge, ways forward might work best by starting with small steps.

Challenges call for systems thinking. The challenges in the Kabul, Kurram and Gomal basins are extensive. The people living in these basins face high poverty levels and are vulnerable to floods and droughts. Climate change will increase temperatures and intensify rainfall variability. Water quality could decline due to increased economic activity and population growth. The data that can be used to make informed decisions is often of variable quantity and quality. Ultimately, aspirations should go beyond addressing immediate crises to fostering transformational changes that provide a platform for human development. To achieve this aim, innovative approaches will be required. Gains that can be achieved by thinking more broadly about the transboundary nature of the overall basin system and cross-sectoral relationships and all water sources need to be considered.

Ways Forward

The Sustainable Development Goals (SDGs)

The outcome of a global multiyear process, the SDGs provide a powerful framework for responding to development challenges across the world. Adopted in late 2015 with a time horizon of 2030, the SDGs have established new goalposts for improving the world's conditions. Consisting of 17 goals and 169 targets, the SDGs have a dedicated water goal, Goal 6, that seeks to 'ensure availability and sustainable management of water and sanitation for all'. The SDGs have driven a large body of work related to the formulation and implementation of Goal 6. The African Ministers' Council on Water, Swiss Confederation, United Nations Global Compact, United Nations University, and UN-Water, for example, have published work on how SDG 6 should be formulated (UNU-UNOSD, 2013; UN Water, 2014). Researchers such as Hajer *et al.* (2015) have begun to examine how to foster the most efficient implementation of Goal 6 and other SDGs.

SDG 6. The SDGs treat water more holistically than did the Millennium Development Goals. Whereas the Millennium Development Goals focus narrowly on water and sanitation, SDG Goal 6 focuses on water and sanitation as well as a broader set of issues surrounding water management. SDG Goal 6 defines six targets:

- 6.1 – By 2030, achieve universal and equitable access to safe and affordable drinking water for all.
- 6.2 – By 2030, achieve access to adequate and equitable sanitation and hygiene for all and end open defecation, paying special attention to the needs of women and girls and those in vulnerable situations.
- 6.3 – By 2030, improve water quality by reducing pollution, eliminating dumping and minimizing the release of hazardous chemicals and materials, halving the proportion of untreated wastewater and substantially increasing recycling and safe reuse globally.
- 6.4 – By 2030, substantially increase water use efficiency across all sectors and ensure

sustainable withdrawals and supply of freshwater to address water scarcity and substantially reduce the number of people suffering from water scarcity.

- 6.5 – By 2030, implement integrated water resources management at all levels, including through transboundary co-operation as appropriate.
- 6.6 – By 2020, protect and restore water-related ecosystems, including mountains, forests, wetlands, rivers, aquifers and lakes.
- 6.6.a – By 2030, expand international co-operation and capacity-building support to developing countries in water- and sanitation-related activities and programmes, including water harvesting, desalination, water efficiency, wastewater treatment, recycling and reuse technologies.
- 6.6.b – Support and strengthen the participation of local communities in improving water and sanitation management.

The Kabul, Kurram and Gomal basins are not advancing rapidly toward SDG 6. While it would be possible to quantify progress towards each target in the three basins, the outcome would be in little doubt. Indicators of human development in the three shared basins reflect a highly challenged human and physical landscape. Access to potable water is low, food insecurity is high and co-operation on water hardly exists. To make progress towards realizing SDG 6, the breadth and severity of current conditions call for major changes to realize a basic platform for socio-economic development.

More than incremental change is needed to achieve a basic level of human development. The level of structural change needed to foster socio-economic development is likely to be significant. Big changes do not come easily. As reflected in the UN 2023 Water Conference at UN Headquarters in New York, 22–24 March 2023, there is a growing recognition of the need to catalyse and accelerate change in the water sector across the world due to insufficient progress in achieving water security. In the global context, Afghan–Pakistani shared waters may present extreme examples of the lack of progress.

Achieving Co-operation Between Countries to Re-engage in the Region

Development is unlikely to be realized in isolation. From an economic perspective, there are at least two preconditions to catalysing the growth needed for human development: stability and investment. Achieving these two preconditions is likely to be achieved through re-engaging at a regional level, and water co-operation can form a key part of this engagement.

Given its volatility, Afghanistan may have a particular incentive to engage with its neighbours to negotiate and conclude regional framework agreements to create a stable environment that promotes development. In this context, it would seem that fostering co-operation on water between Afghanistan and Pakistan would be an urgent and achievable first step.

References

Hajer, M., Nilsson, M., Raworth, K., Bakker, P., Berkhout, F. *et al.* (2015) Beyond cockpit-ism: four insights to enhance the transformative potential of the Sustainable Development Goals. *Sustainability* 7(2), 1651–1660. DOI: 10.3390/su7021651.

UNU (United Nations University) and UNOSD (United Nations Office for Sustainable Development) (2013) *Catalysing Water for Sustainable Development and Growth: Framing Water Within the Post 2015 Development Agenda: Options and Considerations*. United Nations University Institute for Water, Environment and Health, United Nations Office for Sustainable Development, and Stockholm Environment Institute. Available at: https://i.unu.edu/media/ourworld.unu.edu-en/article/8495/Catalyzing-Water.pdf (accessed 9 July 2022).

UN Water (2014) *A Post-2015 Global Goal for Water: Synthesis of Key Findings and Recommendations from UN-Water*. UN-Water, Geneva, Switzerland. Available at: https://www.un.org/waterforlifedecade/pdf/27_01_2014_un-water_paper_on_a_post2015_global_goal_for_water.pdf (accessed 9 July 2022).

Index

Note: **Bold** page numbers refer to tables and *italic* page numbers refer to figures.